Rational Fault Analysis

ELECTRICAL ENGINEERING AND ELECTRONICS

A Series of Reference Books and Textbooks

Editor

Marlin O. Thurston
Department of Electrical Engineering
The Ohio State University
Columbus, Ohio

1. Rational Fault Analysis
   edited by Richard Saeks and S. R. Liberty

Other Volumes in Preparation

Rational Fault Analysis

edited by

Richard Saeks

*Departments of Electrical Engineering
and Mathematics
Texas Tech University
Lubbock, Texas*

Stanley R. Liberty

*Department of Electrical Engineering
Texas Tech University
Lubbock, Texas*

MARCEL DEKKER, INC. New York and Basel

Library of Congress Cataloging in Publication Data

Symposium on Rational Fault Analysis, Texas Tech University, 1974.
  Rational fault analysis.

  (Electrical engineering and electronics ; 1)
  Co-sponsored by the Office of Naval Research and the
Army Research Office.
  Bibliography: p.
  Includes index.
  1. Electronic apparatus and appliances--Testing--
Congresses. 2. Electronic apparatus and appliances--
Reliability--Congresses. 3. Automatic checkout equip-
ment--Congresses. I. Saeks, Richard. II. Liberty,
S. R., 1942-   III. United States. Office of Naval
Research. IV. United States. Army Research Office.
V. Title. VI. Series.
TK7870.S945 1974   621.3815'1'028   76-40363
ISBN 0-8247-6541-9

MARCEL DEKKER, INC.
270 Madison Avenue, New York, New York 10016

Current Printing (last digit):
10 9 8 7 6 5 4 3 2 1

PRINTED IN THE UNITED STATES OF AMERICA

PREFACE

This volume represents the proceedings of the Symposium on Rational Fault Analysis held at Texas Tech University August 19-20, 1974. The purpose of the symposium was to bring together individuals from industry, government and universities interested in the development of rational procedures for the detection, location and prediction of faults in a wide variety of systems to determine the present state-of-the-art and directions for future research. The symposium was organized about a number of presentations outlining the state-of-the-art in various areas. Topics included fault analysis in analog and digital electronics, testing of non-electrical devices, implementation problems, and the mathematics of fault analysis.

We would like to thank the Office of Naval Research and the Army Research Office for their support of the conference and of these proceedings under ONR contract N00014-73A-0434-003 and Mrs. Flores Myers of the Texas Tech Division of Engineering Services for her invaluable aid in preparing the manuscript.

R. Saeks
S. R. Liberty
Co-Chairmen for the
Symposium on Rational Fault Analysis

FOREWORD

A recent DoD study, entitled Electronics-X, pointed out that of the $84 billion DoD budget of FY 74, $15.3 billion was related to the electronics area. Of this, fully $5 billion was allocated to electronics maintenance and consequently to the problem of reliability. Now, with the exception of fault tolerant computing, reliability science has primarily been centered around the test and evaluation shops, and one may rightly ask just why the basic research houses, such as ONR and ARO are butting into this business.

In the recent past, however, two trends in science have led us to rethink the area of fault analysis and the research goals of automatic (built-in) testing. One of these is the rapid advance in the area of system theory and the consequent architectural understanding of high level systems. In the past, approaches have been either local, in which the derivatives of the various external system parameters with respect to component parameters are used (i.e. sensitivity) to achieve an estimate, or they have been global in which simulations of all possible faulty conditions are made prior to the test and compared with the measured external system paramters to achieve the desired estimates. Present system theory techniques have advanced to the point where more meaningful approaches are possible. One approach, is based upon the observation that the connections in a circuit or system define a system valued function of a system valued variable, i.e., a mapping of the internal component parameters to the externally measured system parameters. The strength of this approach is that it allows the utilization of a maximum amount of data about the system. Such techniques have applications to systems ranging from a multi-million dollar AEGIS radar to testing of the smallest IC chip just off the production line. But these in themselves are insufficient to provide the *raison d'etre* for electronics research efforts in this area. Rather, it is the prospect raised by the second scientific trend that will allow the implementation of these techniques and hopefully open the door to rational fault analysis.

Let me now turn to this second scientific advance, namely the development of large scale integrated (LSI) systems and consequently microprocessors that are microprogrammable and cheap. To understand their impact, we must look into our electronics systems. A major portion of the cost of our sophisticated electronics systems today is the central data processor and its consequent hardware. Fully 85% of today's computer costs are related to the software required to make a general purpose processor do complex calculations rapidly. But, more importantly, this impacts on system design, and the hidden costs are much larger. The LSI revolution will allow us to circumvent this problem by reverting to distributed function processing in our electronic systems. This involves a return in our design of tactical systems to allowing design engineers to first analyze the system needs and then choose an optimum dedicated processor to fulfill those needs. Here is where the impact of the LSI microprocessor will be felt, in dedicated micropro-grammable microprocessors for specific computational needs. In particular, the Electronics-X study mentioned above points out two highest priority recommendations that are appropriate to us:

- "System-function-oriented processing-hardware structures should be considered as alternatives to the conventional centralized programmable uniprocessor for use in military tactical systems.
- "A processor design for each system should be selected and developed that will minimize the combined costs of hardware and software; the allocation of functions between hardware, software, and human operators should be consciously worked out prior to decision."

The first of these is already being implemented, for example in the DAIS system under development at the AFAL, and in system development work at the NELC. The second of these recommendations is readily implemented with the advancing LSI technology of today. These approaches relegate the central tactical processor to a simple command and control monitoring function, a function easily carried out by a minicomputer or another microcomputer. In each case, the software is minimal and compatible with hardware implementation.

But, where does fault analysis fit into this distributed function system? The size requirement of a microprocessor is trivial. As soon as you accept distributed function processing, you realize that it is exceedingly trivial to add an additional microprocessor to each subsystem, whose dedicated role is to do *in situ* fault analysis of that subsystem. The implementation of this depends only upon the system theoretic ideas discussed above.

So the confluence of these two areas of scientific endeavor now open the door to true real-time rational fault analysis in our systems, in a predictive mode. It is with these reasons in mind that the Office of Naval Research and the Army Research Office have co-sponsored the Symposium on Rational Fault Analysis and these proceedings.

Dr. D. K. Ferry
Office of Naval Research

CONTRIBUTORS

S. D. Bedrosian, Moore School of Electrical Engineering, University of Pennsylvania, Philadelphia, Pennsylvania

R. F. Garzia, Senior Application Analyst, Babcock and Wilcox Company, Barberton, Ohio

J. R. Greenbaum, Re-entry and Environment Systems division, General Electric Company, Philadelphia, Pennsylvania

S. Louis Hakimi, Department of Electrical Engineering and Computer Sciences, Northwestern University, Evanston, Illinois

John P. Hayes, Department of Electrical Engineering and Computer Science, University of Southern California, Los Angeles, California

S. R. Liberty, Department of Electrical Engineering, Texas Tech University, Lubbock, Texas

Gary K. Maki, Electrical Engineering Department, University of Idaho, Moscow, Idaho

John F. Meyer, Departments of Electrical and Computer Engineering and Computer Science, University of Michigan, Ann Arbor, Michigan

N. N. Puri, Rutgers University, New Brunswick, New Jersey

M. N. Ransom, Bell Laboratories, Naperville, Illinois

Jean-Claude Rault, Thomson-CSF, DIB, Service Recherche, Paris, France

R. Saeks, Departments of Electrical Engineering and Mathematics, Texas Tech University, Lubbock, Texas

Dwight H. Sawin III, Communications/ADP Laboratory, USAECOM, Ft. Monmouth, New Jersey

L. Tung, Department of Electrical Engineering, Texas Tech University, Lubbock, Texas

CONTENTS

Rational Fault Analysis

FAULT ANALYSIS IN DIGITAL SYSTEMS –
A GRAPH THEORETIC APPROACH

S. Louis Hakimi

Introduction

The concept of reliable computation or, in more modern terms, "fault-tolerant computing" has been a serious concern of computer engineers and scientists from the inception of the field of electronic computers. The early researchers basically concerned themselves with various forms of hardware redundancies to improve the reliability [1,2]. More recently the possibility of the automatic repairing of the "component" malfunctions has gained credibility. Of course, such a possibility requires the ability to isolate or to identify the faulty components before the replacement of these components can be contemplated. There has been a great variety of techniques for diagnosing or identifying faults in digital systems. Most of these procedures seem to be somewhat *ad hoc* and often only applicable to the very particular situations for which they were intended.

A serious attempt to view "fault diagnosis" problems abstractly was initiated by Preparata, Metze, and Chien [3], who used an elegant graph theoretic model and who defined the motion of t-diagnosable systems. On first exposure, the reader may find this approach, although enlightening, to be applicable to the situations in which one has a network of mini-computers or a network of microprocessors or possibly to sociological (human interaction) situations. But these misgivings substantially fade when Russel-Kime's [4] generalization of Preparata, Metze, and Chien's model is presented. However, it must be admitted that such a generalization is achieved at the cost of the simplicity and the transparency of the results.

In the next section, we begin by discussing probabilistic fault diagnosis which leads to the notion of probabilistically-t-diagnosable and t-diagnosable systems. We proceed in Section II, by giving characterization theorems (1) for

S. Louis Hakimi is with the Departments of Electrical Engineering and Computer Sciences, Northwestern University, Evanston, Illinois.

This work is supported by the U.S. Air Force Office of Scientific Research, Air Force Systems Command, Grant AFOSR-71-2103, (Amend. C).

1

t-diagnosable systems by Hakimi and Amin [5] and (2) for probabilistically t-diagnosable systems by Maheshwari and Hakimi [6]. And finally, Section II terminates by discussion of a number of unsolved problems. Section III presents Russel-Kime's main result and its generalization by Maheshwari and Hakimi [6].

Fault Analysis in the Preparata–Metze–Chien Model

Consider a simple situation, shown in Fig. 1, where unit u_1 is testing unit u_2. The result of that test, designated by $w(u_1, u_2)$, is the weight associated with the directed line or, using graph-theoretic terminology, the arc (u_1, u_2). $w(u_1, u_2) = 1$, if u_1 finds u_2 to be faulty and $w(u_1, u_2) = 0$ otherwise. If $w(u_1, u_2) = 1$, then the possibility of both u_1 and u_2 being non-faulty must be excluded. If $w(u_1, u_2) = 0$, then the possibility of u_1 being non-faulty while u_2 is faulty must be excluded. Given some *a priori* probabilities p_1 and p_2 for u_1 and u_2 to be faulty and the test outcomes, one can easily find the most likely faulty unit(s). For example, if $p_1 = 0.6$ and $p_2 = 0.4$ and $w(u_1, u_2) = 0$, then it is most likely that u_1 is faulty while u_2 is non-faulty.

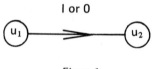

Figure 1

Now consider a more complex situation, depicted in Fig. 2, where we have a "network" of eight units (minicomputers, etc.) which are assigned, in a designated manner, to test each other. The outcomes of the tests are shown as weights on the various arcs corresponding to these tests. Momentarily, let us assume that all units have the same (*a priori*) probabilities of being faulty and this probability is less than ½. Then, what is the most likely set of faulty units? Under these circumstances, it can be shown that $F_1 = \langle u_4, u_6, u_8 \rangle$ is the most likely set of faulty units. More precisely it can be shown that it is possible that all units in F_1 are faulty and the remaining units are non-faulty and there is no other set of units F_2 with the same property which has a cardinality no greater than F_1 (or whose probability of being faulty is no less than that of F_1).

This now brings us to the notion of Consistent Fault Sets (CFS). A subset of units F_1 is a consistent fault set if there is no contradiction in the test outcomes. We shall make this notion more precise. Before that, however, we should note that the "network" which represents the test assignments for the system for the purpose of diagnosis can in general be considered to be a digraph G(U,T) whose vertices U correspond to the units and whose arcs correspond to the tests in the system. If the outcomes of the tests are given as the weights of the arcs, then the resulting digraph is a weighted digraph and is represented by $G_W(U,T)$. The following assertion clarifies the notion of a consistent fault set.

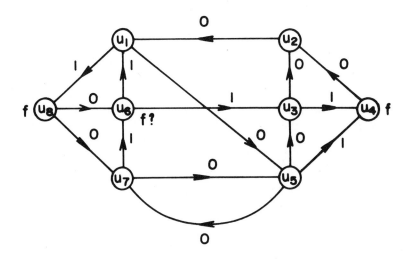

Figure 2

Assertion: Given $G_W(U,T)$, then $F \subseteq U$ is a consistent fault set (CFS) if and only if

(a) u_i and $u_j \in U - F$ and $(u_i,u_j) \in T$ implies $w(u_i,u_j) = 0$

(b) $u_i \in U - F$, $u_j \in F$ and $(u_i,u_j) \in T$ implies $w(u_i,u_j) = 1$

Given $G_W(U,T)$, there are usually many consistent fault sets and U itself is always a CFS. We now give a method for finding the most likely CFS using a $\langle 0,1 \rangle$-programming formulation. To do this, we first assume p_i is the probability of unit u_i being faulty for $i = 1,2,\ldots,n = |U|$. We associate a binary variable x_i to each unit u_i, $i = 1,2,\ldots,n$ with the stipulation that if u_i is faulty $x_i = 1$ and if u_i is non-faulty $x_i = 0$.

Let us consider the following set of $|T|$ linear inequalities:

$x_i + x_j \geqslant 1$, whenever $(u_i,u_j) \in T$ and $w(u_i,u_j) = 1$

$x_i - x_j \geqslant 0$, whenever $(u_i,u_j) \in T$ and $w(u_i,u_j) = 0$

It is easy to see that any binary solution to the system of inequalities corresponds to a CFS and vice-versa, i.e., if $(x_1 x_2 x_3 \cdots x_n) = (101 \cdots 10)$ is a solution to the above inequalities, then $\langle u_1, u_3, \ldots, u_{n-1} \rangle$ is a CFS. It is now clear that to find a most likely CFS, one must have an objective function.

It can be shown that to find a most likely CFS, one must

$$\max \sum_{i=1}^{n} x_i \log(\frac{p_i}{1 - p_i})$$

subject to the above inequality constraints [6]. If $p_1 = p_2 = \cdots = p_n = p < \frac{1}{2}$, then the above objective function becomes

$$\max \sum_{i=1}^{n} -x_i = \min \sum_{i=1}^{n} x_i$$

The above $\langle 0,1 \rangle$-programming problem is intractable in the sense of Karp [7], because it contains the cover problem as a special case. Hence it is highly unlikely that one could find a polynomially bounded algorithm for its solution.

We now define two quantities that are measures of diagnostic capability of a system represented by $G(U,T)$.

Definition 1, [3]: Given a positive integer t, a system $G(U,T)$ is t-diagnosable if for every set of test outcomes there exists at most one consistent fault set F with $|F| \leqslant t$.

Definition 2, [6]: Given $t > 0$, a system $G(U,T)$ is probabilistically-t-diagnosable (pr-t-diagnosable) if for every set of test outcomes, there exists at most one consistent fault set F with

$$\Pr\langle F \text{ being the faulty set} \rangle = \prod_{u_i \in F} p_i \prod_{u_i \in F} (1 - p_i) > t$$

It should be noted that in a t-diagnosable system, if one could exclude the possibility of having more than t faulty units, then one can uniquely identify the faulty units (correctly diagnose the system faults) given $G_w(U,T)$. The notion of t-diagnosability is very much related to the concept of Hamming distance d $(= 2t + 1)$ in algebraic coding theory [8]. (We shall say more about this relation later.) On the other hand, in a pr-t-diagnosable system, if one could stipulate that only those fault sets could occur that have *a priori* probability of occurrence which is larger than t, then one can correctly identify the faulty units.

A measure of reliability of a system is the probability p_c of correctly identifying the faulty units. It is easy to see that in a t-diagnosable system in which $p_1 = p_2 = \ldots = p_n = p$

$$p_c \geqslant \sum_{i=0}^{t} \binom{n}{i} p^i (1-p)^{n-i}$$

and in a pr-t-diagnosable system

$$p_c \geqslant \sum_{F \in \langle S \subset U \mid Pr(S) > t \rangle} Pr(F)$$

where for each $S \subset U$, $Pr(S) = \prod_{u_i \in S} p_i \prod_{u_i \in U-S} (1 - p_i)$, and the decision of

the choice of CFS is based on the most likelihood decision scheme.

We now give a characterization of t-diagnosable systems. Given $G(U,T)$, by $d_{in}(u)$, called in-degree of $u \in U$, we mean the cardinality of the set $\langle u' \in U \mid (u', u) \in T \rangle$. For $S \subseteq U$, we define $\Gamma S = \langle \cup \Gamma u \rangle - S u \in S$ where for $u \in U$, $\Gamma u = \langle u' \mid u \in S \in U \mid (u,u') \in T \rangle$.

Theorem 1 (Hakimi and Amin [5]): A system $G(U,T)$ with $|U| = n$ is t-diagnosable if and only if

(a) $\qquad n \geqslant 2t + 1$

(b) $\qquad d_{in}(u) \geqslant t, \; u \in U$

(c) $\qquad$ for every $S \subset U$ with $n - 2t \leqslant |S| < n - t$, we have
$\qquad |\Gamma S| > |S| - n + 2t$

Remarks: The necessity of (a) and (b) was established by Preparata, Metze and Chien [3]. The necessity of (c) and the sufficiency of all three conditions together were established by Hakimi and Amin [5]. It may be worth noting that (a) is implied by (c); that is, condition (a) is redundant.

It is quite time consuming to test a system for t-diagnosability, because to test whether or not condition (c) is satisfied one must examine a large class of subsets of U. Hakimi and Amin [5] discovered another theorem which gives an easy characterization of the class of t-diagnosable systems in which no two units are allowed to test each other.

Theorem 2 (Hakimi and Amin [5]): Let $G(U,T)$ represent a system in which no two units test each other. Then $G(U,T)$ is t-diagnosable if and only if $d_{in}(u) \geqslant t$ for all $u \in U$.

Not only does the condition of Theorem 2 provide a simple test for t-diagnosability, but also it can be conveniently used for design of "least cost" t-diagnosable systems. To see this, consider a situation in which we want to exclude the possibility of any two units testing each other and we are given the cost c_{ij} of unit u_i to test unit u_j for all i and j. Then one can find a least cost t-diagnosable system in which no two units test each other, using a graph-theoretic polynomially bounded algorithm. The interested reader is referred to Hakimi, Maheshwari and Amin [9]. The problem of least cost design of t-diagnosable systems, when pairs of units are allowed to test each other, seems much more difficult.

We now consider the characterization of pr-t-diagnosable systems. As before, let p_i be the probability of unit u_i being faulty and let $Pr(F)$ be the probability of F being the faulty set, i.e., $Pr(F) = \prod_{u_i \in F} p_i \prod_{u_i \in U\text{-}F} (1\text{-}p_i)$.

It can be seen that $Pr(F) > t$ is equivalent to

$$\sum_{u_i \in F} \log \frac{1 - p_i}{p_i} < K(t)$$

where $K(t) = -\log t + \sum_{i=i}^{n} \log (1 - p_i)$. Therefore, if we associate to each unit u_i, the weight $w(u_i) = \log \frac{1 - p_i}{p_i}$, then $Pr(F) > t$ if and only if $\sum_{u_i \in F} w(u_i) < K(t)$.

We now have the following necessary condition for pr-t-diagnosability.

Theorem 3 (Maheshwari and Hakimi [6[): If $G(U,T)$ corresponds to a pr-t-diagnosable, then

$$w(u_i) + \sum_{u_j \in \Gamma^{-1} u_i} w(u_j) \geqslant K(t) \ , \quad \text{for all } u_i \in U$$

where $\Gamma^{-1} u_i = \langle u_k \in U \mid (u_k, u_i) \in T \rangle$.

Proof: Suppose otherwise, that is, there exists $u_r \in U$ such that

$$w(u_r) + \sum_{u_j \in \Gamma^{-1} u_r} w(u_j) < k(t)$$

Consider the following set of test outcomes: $w(u_i,u_j) = 1$ whenever $u_i \in U$, $u_j \in \Gamma^{-1}u_r$ and $(u_i,u_j) \in T$ and $w(u_i,u_j) = 0$ otherwise. Then it is easy to see that $\Gamma^{-1}u_r$ and $\Gamma^{-1}u_r \cup \langle u_r \rangle$ are both consistent fault sets whose probability of occurrence is greater than t. Hence the theorem.

Theorem 4 (Maheshwari and Hakimi [6]): Let $G(U,T)$ correspond to a system in which no two units test each other. Then $G(U,T)$ is pr-t-diagnosable if

$$\tfrac{1}{2}w(u_i) + \sum_{u_j \in \Gamma^{-1} u_i} w(u_j) \geqslant K(t) , \quad \text{for all } u_i \in U$$

and $p_i < \tfrac{1}{2}$ for $i = 1, 2, \ldots , n$.

The proof of Theorem 4 is a bit long and is based on the duality theory of linear programming. Note that the sufficient condition of Theorem 4 is very close to the necessary condition given in Theorem 3. But it can be shown that neither the condition of Theorem 3 is sufficient (even for the case when no two units test each other) nor the condition of Theorem 4 is necessary for pr-t-diagnosability [9,10].

A completely general characterization of pr-t-diagnosability is quite complex and the interested reader can find such a result in Maheshwari and Hakimi [6]. Even the statement of the theorem would require a number of preliminary definitions and thus will not be given here.

We shall now turn our attention to the sequentially-t-diagnosable (s-t-diagnosable) system which was also introduced by Preparata, Metze, and Chien [3]. They defined a s-t-diagnosable system as a system in which given $G_w(U,T)$ and that $1 \leqslant$ number of faults $\leqslant t$, then one can identify at least one faulty unit correctly. It is understood that once one identifies one faulty unit, then one repairs that unit and proceeds to "sequentially" identify the other faulty units. (It should be noted that if the value of all tests is "0", then one can assume there is no fault.) Given the number of units n and positive integer t with $n > 2t$, Preparata, Metze, and Chien [3] proved that there exists a s-t-diagnosable system with n + 2t - 1 tests. The following theorem is a significant improvement of this result.

Theorem 5 (Maheshwari and Hakimi [6]): Given the number of units n and a positive integer t with $n > 2t$, there exists a s-t-diagnosable system with n + t - 1 tests.

We will not present a complete proof of Theorem 5 here. However, we will give the test assignments in such a s-t-diagnosable system as well as a procedure for identifying a faulty unit. The test assignment, as usual, is represented by a digraph denoted by $G(U,T)$. Let $U = \langle u_0, u_1, \ldots, u_{n-1} \rangle$ corresponding to the set of units and $T = T_1 \cup T_2$ corresponding to the set of tests, where $T_1 = \langle (u_0, u_1), (u_1, u_2), (u_2, u_0), \ldots, (u_0, u_{2i-1}), (u_{2i-1}, u_{2i}),$

$(u_{2i}, u_0), \ldots, (u_0, u_{2t-1}), (u_{2t-1}, u_{2t}), (u_{2t}, u_0) \rangle$ and $T_2 = \langle (u_0, u_{2t+1}), (u_0, u_{2t+2}), \ldots, (u_0, u_{n-1}) \rangle$. Such a digraph is shown in Fig. 3. As it can be seen the arcs in T_1 consist of t directed 3-cycles and the remaining arcs correspond to the tests performed by u_0 on remaining units $u_{2t+1} \ldots u_{n-1}$. The results of the tests in each 3-cycle, say $(u_0, u_{2i-1}) (u_{2i-1}, u_{2i}) (u_{2i}, u_0)$, is represented by binary 3-sequency $w(u_0,u_{2i-1})w(u_{2i-1},u_{2i})w(u_{2i},u_0)$. We shall call the 3-sequence $w(u_0,u_{2i-1})w(u_{2i-1},u_{2i})w(u_{2i},u_0)$ the test pattern corresponding to its 3-cycle. If the test pattern corresponding to any cycle contains only one "1", then the unit corresponding to the "head" of the arc whose weight is equal to "1" is identified as a faulty unit (see Theorem 6). Thus, we can assume none of the 3-cycles has a test pattern which is a permutation of the sequence 001. Let n_1, n_2, n_3, n_4, and n_5 be the number of cycles with test patterns 000, 111, 110, 101, and 011, respectively. Now it can be shown that [6] if $1 + 2n_1 + n_2 + n_3 + n_5 \leqslant t$, then u_0 is faulty and otherwise u_0 is non-faulty. If u_0 is non-faulty, then since u_0 tests other units, at least one other faulty unit will be identified.

  The s-t-diagnosable systems are extremely interesting because: (1) they require far fewer tests (n + t - 1) than t-diagnosable systems, which require

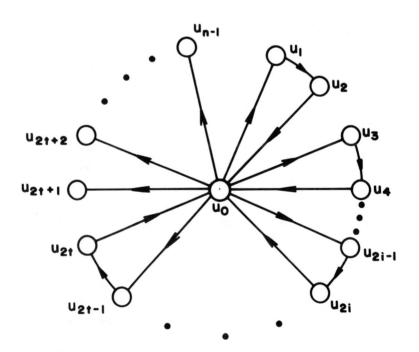

Figure 3

(nt) tests, and (2) as was described above, there is a very simple algorithm for identifying at least one faulty unit. However, we do not have as yet a complete characterization of s-t-diagnosable systems. Such a characterization seems to be very difficult. For further generalizations of s-t-diagnosable systems, the interested reader is referred to Maheshwari [10].

Although the problem of finding the most likely (or the least cardinality) consistent fault set is intractable, it is fairly easy to construct a polynomially bounded algorithm to find the CFS of least cardinality in a t-diagnosable system when the number of faults are no larger than t [9]. A very efficient algorithm, say linear, however, would be welcome.

One may ask: given $G_w(U,T)$, what does it take to find one unit that is faulty? Note that in contrast with previous paragraphs, no assumption as to the number of faulty units is made here. The following theorem completely answers this question. We first define 0-path in $G_w(U,T)$ to be a directed path of length zero or a directed path in which all arcs have weight zero where the length of the path is the number of arcs in it.

Theorem 6 (Hakimi, Maheshwari, and Amin [9]): Given $G_w(U,T)$ a unit u_i can be identified to be faulty if and only if there exists two 0-paths starting at u_i and ending at u_r and u_s such that $(u_r,u_s) \in T$ and $w(u_r,u_s) = 1$.

We will discuss one other issue of considerable importance before we close this section. We stated earlier in this section that in a t-diagnosable system in which $p_1 = p_2 = \cdots = p_n = p$, we have

$$p_c \geqslant \sum_{i=0}^{t} \binom{n}{i} p^i(1-p)^{n-i}$$

Of course, it is assumed that one always uses the most likely decision scheme, i.e., one picks the CFS of the least cardinality given $G_w(U,T)$. As in the case of algebraic coding theory, there are a great many other sets of faults whose cardinality is larger than t which could be uniquely identified using such a decision scheme. Thus, in many cases, one has

$$p_c \gg \sum_{i=0}^{t} \binom{n}{i} p^i(1-p)^{n-i}$$

As in coding theory, however, we do not have an algorithm for finding the unique consistent fault set (of the least cardinality when it exists) when the number of faults exceeds t. Also we have no idea about test structures that increase p_c. There is no parallel concept here to the notions of perfect or pseudo-perfect codes in coding theory [8] which are the codes which have a given Hamming distance equal to the largest possible probability of correct decoding.

Fault Analysis in Russell-Kime Model

Suppose we have a set of m tests $T = \langle t_1, t_2, \ldots, t_m \rangle$ and a set of n units (modules, potential faults, etc.) $U = \langle u_1, u_2, \ldots, u_n \rangle$. Let us consider a digraph $G(U,T;A)$ where $U \cup T$ is the vertex set of the digraph and A is its arc set. Each arc in A is either of the form (u_i, t_j) or (t_r, u_s), i.e., there are no arcs between a pair of vertices in U or between a pair of vertices in T. Furthermore, we assume that if for some $u_i \in U$ and $t_j \in T$, $(u_i, t_j) \in A$, then $(t_j, u_i) \notin A$ and if similarly $(t_j, u_i) \in A$ implies that $(u_i, t_j) \notin A$. Mappings $\Lambda : T \rightarrow U$ and $\Omega : U \rightarrow T$ are defined as follows: $\Lambda t_j = \langle u_j \in U \mid (t_j, u_j) \in A \rangle$ and $\Omega u_i = \langle t_i \in T \mid (u_i, t_k) \in A \rangle$. For $T' \subseteq T$, $\Lambda T' = \bigcup_{t_i \in T'} \Lambda t_i$ and similarly one defines $\Omega U'$

for $U' \subseteq U$. Finally for $u_i \in U$, $\Lambda^{-1} u_i = \langle t_j \in T \mid (t_j, u_i) \in A \rangle$ and similarly $\Omega^{-1} t_k$ can be defined. Definitions $\Lambda^{-1} U'$ and $\Omega^{-1} T'$ follow in self-explanatory manners.

For the digraph $G(U,T;A)$ to represent a model of a diagnosable system, we must associate some significance to the outcome $w(t_i) \in \langle 0,1 \rangle$ of each test $t_i \in T$. For this purpose, we assume that $w(t_i) = 1$ implies that at least one unit in $\Lambda t_i \cup \Omega^{-1} t_i$ is faulty and $w(t_i) = 0$ implies that if $\Omega^{-1} t_i$ has no faulty units then neither does Λt_i. One can interpret Λt_i to be the set of units which t_i could potentially "diagnose" and $\Omega^{-1} t_i$ is the set of units which, when containing a faulty unit, invalidates the result of test t_i.

We now define the notion of consistent fault set (CFS) in the Russell-Kime Model. Given $G(u,T;A)$ and the test results, a set $F_1 \subseteq U$ is a CFS if and only if [6]

(a) $w(t_i) = 1$, for all $t_i \in \Lambda^{-1} F_1 - \Omega F_1$
(b) $w(t_i) = 0$, for all $t_i \in T - (\Lambda^{-1} F_1 \cup \Omega F_1)$

As before if probability of u_i being faulty is p_i for $i = 1, 2, \ldots, n$, then one can ask the question: given $G(U,T;A)$ and test outcomes, what is the most likely consistent fault set? The problem can be formulated as a $\langle 0\text{-}1 \rangle$-programming problem as follows. Let x_i be a binary variable associated with unit u_i with the stipulation that u_i is faulty if $x_i = 1$ and vice versa. The formulation is

$$\text{maximize} \quad \sum_{i=1}^{n} x_i \, \log \frac{p_i}{1 - p_i}$$

subject to

$$\begin{cases} \sum_{u_k \in \Lambda t_i} x_k + \sum_{u_j \in \Omega^{-1} t_i} x_j \geq 1, \text{whenever } w(t_i) = 1 \\ \\ -\sum_{u_k \in \Lambda t_i} x_k + |\Lambda t_i| \sum_{u_j \in \Omega^{-1} t_i} x_j \geq 0 \text{ , whenever } w(t_i) = 0 \end{cases}$$

It is instructive to note that the Russell-Kime Model reduces to the Pre-parata-Metze-Chien Model when $|\Omega^{-1} t_i| = 1$ and $|\Lambda t_i| = 1$ for $i = 1,2,\ldots,m$.

A system represented by digraph $G(U,T;A)$ is t-diagnosable for some integer $t > 0$, if for every set of test outcomes there exists at most one consistent fault set $F \subset U$ with $|F| \leqslant t$. Let $F_1 \subseteq U, F_1$ *masks* $u_i \in U$ if $\Lambda^{-1} u_i \subseteq \Omega F_1$. The masking index of u_i, denoted by $m(u_i)$, is the cardinality of smallest subset of U which masks u_i. For $F_1 \subseteq U$, let $C(F_1) = \Lambda^{-1} F_1 - \Omega F_1$ and $E(F_1) = F_1 \cap \langle \Lambda C(F_1) \rangle$. $E(F_1)$ can be interpreted to be the subset of F_1 for which there remains a valid test in the presence of F_1.

Theorem 7 (Russell and Kime [4]): A system represented by digraph $G(U,T;A)$ with $|\Lambda t_i| = 1$ for all $t_i \in T$ is t-diagnosable if and only if

(a) $m(u_i) \geqslant t, \quad u_i \in U$

(b) for every $F \subset$ with $t < |F| \leqslant 2t, |E(F)| > 2t - |F|$

A more general theorem which gives characterization of t-diagnosable systems, when $|\Lambda t_i| \neq 1$ for at least one $t_i \in T$, is available [6]. However, the result is very complex. For a discussion of the probabilistically-t-diagnosable system in the Russell-Kime Model one is referred to Maheshwari [10]. It is worth noting that Theorem 7 implies Theorem 1 and it was discovered later than Theorem 1 but independently.

In conclusion, this author feels that such abstract and somewhat gross models as considered here lead to extremely enlightening results in fault analysis of digital systems. In contrast with much of the literature, here we have chosen to ignore questions as to how the tests are carried out or how does unit u_i test u_j, in the hope that one can develop a clearer understanding of what to do when the results of the tests are available. As one can see there still remains much to be done. As an example one might want to know how much "information" the result of each new test reveals to us. Such Information Theoretic concepts should play important roles in fault analysis.

REFERENCES

[1] J. von Neumann, "Probabilistic Logics and the Synthesis of Reliable Organisms from Unreliable Components," in *Automata Studies*, C. E. Shannon and J.

McCarthy, Eds., Annals of Mathematics Studies, No. 34, Princeton University Press, Princeton, N.J., 1956.

[2]   E. F. Moore and C. E. Shannon, "Reliable Circuits Using Less Reliable Relays," *J. Franklin Institute*, Part I, Vol. 262, pp. 191-208, Sept. 1956; Part II, pp. 281-297, Oct. 1956.

[3]   F. P. Preparata, G. Metze, and R. T. Chien, "On the Connection Assignment Problem of Diagnosable Systems," *IEEE Trans. on Electronic Computers*, Vol. EC-16, pp. 848-854, Dec. 1967.

[4]   J. D. Russell and C. R. Kime, "On the Diagnosability of Digital Systems," in *Digest of 1973 Int. Symp. on Fault-Tolerant Computing*, IEEE Computer Society Publications, pp. 139-144, June 1973.

[5]   S. L. Hakimi and A. T. Amin, "Characterization of Connection Assignment of Diagnosable Systems," *IEEE Trans. on Computers*, pp. 86-88, Jan. 1974.

[6]   S. N. Maheshwari and S. L. Hakimi, "On Models of Diagnosable Systems and Probabilistic Fault Diagnosis," submitted for publication (July 1974), *IEEE Trans. of Computers*.

[7]   R. M. Karp, "Reducibility Among Combinatorial Problems," in *Complexity of Computer Computations*, R. E. Miller and J. W. Thatcher, Eds., Plenum Press, 1972.

[8]   W. W. Peterson and W. J. Weldon, Jr., Error-Correcting Codes, 2nd Edition, MIT Press, Cambridge, Mass., 1972.

[9]   S. L. Hakimi, S. N. Maheshwari, and A. T. Amin, "On a Graph Model of Diagnosable Systems," in *Modeling and Simulation*,Proc. of 4th Annual Pittsburgh Conference, Vol. 4, pp. 220-223, April 1973.

[10]   S. N. Maheshwari, "Graph-Theoretic Models for Diagnosis of Digital Systems," Ph.D. Dissertation, Dept. of Computer Sciences, Northwestern University, Evanston, Ill., Aug. 1974.

FAULT ANALYSIS OF ANALOG CIRCUITS USING GRAPH THEORY

S. D. Bedrosian

Introduction

Since the area of fault analysis is likely to be new to many of you, a brief historical overview of fault isolation for analog circuits is in order. It must be emphasized that this problem area is in fact *not* new. This is not too surprising if we consider some of the major events in electronics in the past four decades as shown in Figure 1. Prior to WW II there simply was not much

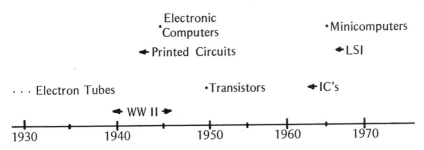

Figure 1. Some milestones in electronics.

electronic equipment around to worry about. Furthermore, with an ordinary AC-DC superheterodyne receiver, for example, one could do meaningful signal tracing to isolate the faulty section simply by touching the grid caps of the electron tubes. That, combined with the fact that about 75% of the faults occurred in the tube filaments, power supply, power cord, and fuses made the maintenance problems manageable.

The sharp increase in electronic usage (avionics, radar, communication, etc.) during WW II focused attention on the critical problem. The existing manual techniques, so highly dependent on the technician's experience and skill, were generally inadequate. The growing complexity as well as the sheer numbers further underscored the lack of systematic approaches. There was, furthermore, no hope of ever having enough experienced technicians. In the

S. D. Bedrosian is with the Moore School of Electrical Engineering, University of Pennsylvania, Philadelphia, Pa. 19174.

following decade large efforts got underway on missiles, weapons systems, jet aircraft, and finally the space program. These new requirements further strained the existing art of fault diagnosis and stimulated some studies.

In spite of these kaleidoscopic events one should note that the fault analysis problem had not really gone unnoticed. This was especially true in the telephone industry and particularly at the Bell Telephone Laboratories. Careful sifting of the literature reveals awareness even though spectacular successes were not being reported. A few examples will suffice. In 1911, Campbell [1] introduced the concept of elimination of concealed circuits by observing that in general one can divide a network into a concealed and an accessible part. He indicated that it is convenient during analysis to eliminate the former from explicit appearance in the impedance determinant when we are concerned only with the effects which are produced in the accessible part of the network due to causes which are also confined to this part of the network. Campbell made an additional contribution in 1922 when he defined "direct capacity," now denoted as capacitance, as

> "... The direct capacities of an electrical system with n given accessible terminals are defined as the n(n-1)/2 capacities which, connected between each pair of terminals, will be the exact equivalent of the system in its external reaction upon any other electrical system with which it is associated only by conductive connections through the accessible terminals."

He also indicated a method of measurement for determining all the values.

We mention in passing the exhaustive treatment bearing on enumeration of possible measurements on linear passive networks with n-accessible terminals given by Riordan [2]. Because of the special transfer definitions used, he obtained 1.3×10^8 measurements for $n = 10$, i.e., approximately one for each person then living in the U.S.A.! The astronomical numbers thus obtained reveal that the difficulty here is not lack of information but an overabundance of it. It is easy to write down many more equations than the number of unknowns appearing in the network. The difficulty then is to establish a way of obtaining a linearly independent set of n(n-1)/2 members which constitute a complete set that will specify the observable behavior of a passive network.

In 1952 McMillan [3] stated that the practical situation of an amplifier chassis from which the electron tubes have been removed was the motivation for his realizability theory, wherein the passive multiterminal network structure is viewed from pairs of accessible terminals. He noted the remaining need for a more adequate theory for such multiterminal networks as follows:

> "... Here, however, nature has been less kind, in that no uniquely simple method is available for describing the performance of such devices as viewed from their terminals. Indeed, basic network

theory has been remiss here, in not even making available a mode of description which is generally applicable — whether simple or not."

By the late fifties some serious efforts were funded to seek a more adequate theoretical basis for fault analysis. There was sufficient progress in the desired transition from art to science to justify seminars, e.g. Battelle Institute 1962 [A1] and short courses at New York University [C1] as well as at the University of Pennsylvania 1965 [C2]. Meanwhile, as noted in Table 1, [B6] various diagnostic test equipments were being produced by several manufacturers. Hence, we can indicate the evolution of maintenance procedures leading to automatic test equipment (ATE) [B1-B9] as

    1. manual techniques,
    2. serially-programmed ATE, and
    3. computer-controlled ATE.

It should be clear that the second category of ATE's is designed for specific system applications. The basic limitations were the motor skills of the operators and the versatility of the multitude of adapters required at the interface with the equipment undergoing test. Note that initially "automatic testing" has referred to the computer implementation of a manually derived test procedure. Thus it primarily replaced the hands of the skilled technician. For the third category, the design of system logic was based on the recognition of the inherent differences between a skilled technician and various types of computers. In addition to basic limitations, inherent in given general-purpose or special-purpose computers, was (and still is) the adequacy of mathematical theory pertinent to maintenance of complex equipment. Successful high speed operation is contingent upon availability of carefully designed suitable algorithmic or heuristic test routines.

All such complex and expensive ATE development stems from some prior consideration of and implicit or explicit decisions regarding mainte-nance philosophy to be pursued. These considerations include: the level of maintenance, question of replacement or repair, where and how fast the action is to be taken, and by whom.

By now the reader should have sensed the fact that the practical exigencies of the system and hardware developments to meet defense or other requirements have led to ATE's listed in Table I without waiting for complete resolution of the difficulties indicated above. In this respect it should be mentioned that in the case of digital electronic systems there has been considerably more progress. See other papers in these proceedings for details on fault analysis of digital circuits as well as Reference 4.

TABLE I

**I. The growth of automatic testing for military systems
is listed here both chronologically and by program sequencers (boldface)** [1]

Name	Manufacturer	Comparator	Service and Application	Year Delivered	Special Features
Motorized cam					
AN/DSM-2	Hycon	Analog	Navy (Terrier)	1952	Used on guidance systems.
DATS	RCA	Analog	AF (airborne fire control)	1958	Uses a dynamic closed loop test, simulates attack courses, and provides as a direct readout a quantitative measure of operational capability.
Stepping switch					
AN/DPM-7	Raytheon	Analog	Navy (Sparrow)	1958	Quantitative dynamic tests on guidance and control by simulation, go/no-go, output recorded.
MK 409 Mod 0	USN/AF	——	Navy (Sidewinder)	1964	Electromechanical, go/no-go via panel lamp indicators.
Punched tape					
SCAT	Sperry Canada	Analog	Navy (Polaris)	1960	Pneumatic-controlled tape reader using block reading technique.
EPEC	Emerson	Digital	Prototype	1963	Used integrated circuitry.
GPATS (AN/GSM-204(V))	Emerson	Digital	AF	1967	Building block concept, includes manual tape simulator for developing new routines (later version available with computer-control for arithmetic and diagnostic testing).
Magnetic tape					
RADFAC	Republic Aviation	Analog	AF (F-105)	1958	Mobile unit, 90-second test time, comparison via cockpit instruments against voice instructions, self-test.
DEE	RCA	Digital	Army Electronics Command (depot)	1961	First U.S. Army ATE to perform computer-controlled communications electronics testing to 100 MHz.
VAST (AN/USM-247)	PRD	Digital	Navy (avionics electronics)	1972	Building block construction, multiple independent test stations time-shared by a central computer, self-check capability and microintegrated circuitry.
Punched card					
13A0140	Bendix Support	Analog	Airlines	1959	Rotary stepping switches for switching functions, lights and meters for readout.
MAPCHE	RCA	Analog/ digital	AF (Atlas)	1960	Mobile version of APCHE, includes analog functions and pneumatic pressure tests.
Magnetic drum					
ACRE	Lockheed	Digital	Navy (Polaris)	1961	Computerized, 1800 r/min drum with 15 360 word capacity, flexowriter printer.
MTE	RCA	Digital	Army (missile systems)	1964	Mobile shelters, pneumatic/hydraulic electronic test capabilities.
Pinboard (patchboard)					
DY-2010	Dymec	None	General	1963	Programming via diode pins on pinboard, outputs on tape.
Magnetic disk					
DDP-224	Computer Control	Digital	NASA (Saturn V)	1964	General-purpose computer used for Saturn V launch display, fast response.
GE-225	General Electric	Digital	NASA	1964	General-purpose computer used for data acquisition, fast response.
Multiprogramming (combinations of punched tape, punched card, magnetic tape, patchboard, keyboard, etc.)					
ACE-SC	General Electric	Analog/ digital	NASA (Apollo)	1964	Multiple-computer installation, real-time go/no-go, continuous sampling, manual/ semiautomatic/automatic modes, high speed, self-test.
DIMATE	RCA	Digital	Army (depot-vehicular)	1965	Computer-controlled, high speed, go/no-go, building block concept, self-test.
AN/GYK-11 (CENPAC)	Burroughs	Digital	Air Force (general purpose)	1967	Controls up to 10 satellite ATE stations, modular construction.

From *IEEE Spectrum*, Aug. 1974. Used by permission.

The Solvability Concept for Fault Diagnosis

The use of graph models in dealing with analog circuits goes back to the pioneering work of Kirchhoff in 1847 [26]. Our motivation here is related to the concept of solvability for fault diagnosis introduced by Berkowitz about 1960 [5].

Given the schematic configuration of an electrical network having a linear graph representation and only a limited number of accessible terminals, what are the conditions for analytical determination of the element values in terms of measurable responses to applicable stimuli?[†] In the case of fault diagnosis, solvability means that the equations representing permissible measurements on the network have unique non-negative solutions yielding the values of all the unknown elements of the given network.

TABLE II

RELATIONSHIP OF SOLVABILITY TO LINEAR ANALYSIS AND SYNTHESIS

Category	Given	Required	Solution
Analysis	Excitation-Network	Response	Always unique
Solvability	Choice of Excitations } Graph { Corresponding Responses	Element Values	Sometimes unique
Synthesis	Excitation-Response	Network	Rarely unique

It should be clear that this circuit-oriented approach represents an effort to establish bounds on diagnosis under ideal conditions. As indicated in Table II, solvability can be considered to be a separate category from analysis or synthesis in linear network theory. This concept yields a necessary condition for fault diagnosis with absolute certainty.

For a network under investigation, we can assume as known the linear graph or schematic configuration and the fact that it contains a limited number of accessible or partially accessible terminals. Consideration is given to the problem where the types of circuit elements are known but their values are unknown initially. The latter are to be determined, if possible, by suitable application of stimuli and suitable measurements at the accessible terminals.

The purpose is to obtain general theoretical conditions for network element-value solvability [7]. For simplicity, only linear networks which

---

[†] Seshu and Waxman [6] note that in contrast to this circuit-oriented point of view, there is the function-oriented point of view to the checkout problem. Since the latter is basically a black box approach, there is no fault isolation capability.

consist of passive constant-parameter bilateral elements are considered. Three types of nodes are permitted in the network:

1.   accessible nodes, A (the usual external terminals),

2.   partially accessible nodes, P (terminals restricted to application or measurement of voltage), and

3.   inaccessible nodes, I (internal or "concealed" nodes).

Then the total number of nodes N is given by their sum as indicated in Figure 2.

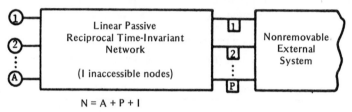

$$N = A + P + I$$

Figure 2. General network for solvability considerations

The observable behavior at the terminals of the network shown in Figure 2 is completely specified by a set of linear algebraic equations. Let Q be designated as the number of admittance functions which specify the network. For a single-element-kind network this is shown to be given by[†]

$$Q = A(A + 2P\text{-}1)/2 \qquad (1)$$

where A and P are the number of accessible and partially accessible nodes respectively.

Formulation of Pertinent Equations and the Key Subgraph

The basic matrix equation

$$I = [Y]V \qquad (2)$$

[†] Note that when P = O, (1) reduces, as it should, to Campbell's result mentioned earlier. Thus, the AP term in (1) accounts for all allowable measurements between the P and A nodes.

represents a set of linear equations describing the steady-state network response. We seek a general procedure for determining element values in the network. The indefinite admittance matrix (IAM) provides a straightforward means for obtaining the transfer admittance parameters for describing the external behavior of a network in terms of the element values. Note, however, that when the element values are considered as the unknowns and the expressions representing short circuit measurements describing its behavior are taken as known constants, the relations obtained are in fact a nonlinear system of equations.

By labeling the N nodes in a given network as shown above, we can write the indefinite admittance matrix as a "compound" matrix (using subscript t for the transpose).

nodes	A	P	I	
A	Y^A	Y^{AP}	Y^{AI}	(3)
P	Y_t^{AP}	Y^P	Y^{PI}	=M
I	Y_t^{AI}	Y_t^{PI}	Y^I	

This M matrix provides a direct method for obtaining a "complete set" of parameters representing short-circuit measurements for describing the behavior from the accessible terminals of the network. The set of linear algebraic equations can be represented in matrix form as

$$Y^A = Y^A E^A + Y^{AP} E^P + Y^{AI} E^I$$
$$I^P = Y_t^{AP} E^A + Y^P E^P + Y^{PI} E^I \qquad (4)$$
$$0 = Y_t^{AI} E^A + Y_t^{PI} E^P + Y^I E^I$$

Then solving for the current at the accessible nodes, we have

$$I^A = Y^{A'} E^A + Y^{P'} E^P \qquad (5)$$

where

$$\left[Y^{A'} \right] = Y^A - Y^{AI} (Y^I)^{-1} Y_t^{AI} \qquad (6)$$
$$\left[Y^{P'} \right] = Y^{AP} - Y^{AI} (Y^I)^{-1} Y_t^{PI} \qquad (7)$$

If there are no partially accessible nodes in the given network, (7) vanishes and (6) simplifies to Kron's reduction formula. Thus, if $P \neq 0$ we have both terms in (5) and the additional equation (7). The result given above can be considered as a generalization of Kron's work. If extensive use will be made of (6) and (7) it is worth introducing some simplifying notation for the common matrix products.

The M matrix representation leads directly to the Q test (1), i.e. $B \leqslant Q$ is a necessary condition for solvability of a single-element-kind network having B branches. The counterpart to the topological test (1) is

$$Q = \begin{pmatrix} \text{Number of possible} \\ \text{off-diagonal elements in } Y^A \end{pmatrix} + \begin{pmatrix} \text{Number of possible} \\ \text{elements in } Y^{AP} \end{pmatrix} \quad (8)$$

A useful check on setting up the compound matrix representation of a network is given by the following theorem: [D2].

THEOREM 1

A set of necessary conditions for solvability in terms of the indefinite admittance matrix **M**, (3), for the given network are:

a. no off-diagonal elements of the matrix contain more than a single term,

b. the diagonal elements of the Y^I submatrix must contain ≥ 3 terms each,

c. the Y^P submatrix must be diagonal, i.e., no off-diagonal elements, and

d. any B_f branches of the network, not incident on the inaccessible nodes (1), will appear only in the submatrix Y^A, or Y^{AP} and Y_t^{AP}.

The proof of parts a, b, and c follows directly from theorems previously given by Berkowitz [7]. Part d warrants special discussion and the introduction of a new concept. It was observed that the entries in the Y^I submatrix, considered from a topological point of view, represent a special subgraph of the given network. This subgraph consists of the subset B_k of all branches incident on all of the inaccessible nodes. Because of its special importance, it is defined as the "key subgraph" G_k of the network. This is indicated in Figure 3 wherein solid lines are the B_k branches and the rest are shown by broken lines. The latter, called the B_f branches, comprise the complementary set of branches of the network.

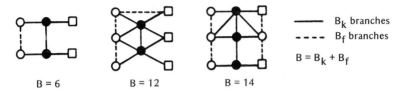

———	B_k branches
- - - -	B_f branches

$$B = B_k + B_f$$

B = 6   B = 12   B = 14

Figure 3. Solvable networks emphasizing the key subgraph

In the context of solvability, identification of the key subgraph leads directly to a set of equations which in the case of a single-element-kind network is a   homogeneous multilinear algebraic form. These equations are modified for convenience in manipulation by using the determinant of the Y^I submatrix as a   scale factor. Details regarding formulation of such equations are shown in the references by Bedrosian and Berkowitz [8,9]. Note in particular the novel formulation of the system of equations that utilizes all of the off-diagonal elements of the "Accessible" portion of the partitioned matrix M rather than the conventional approach of selecting an arbitrary reference node. Thus only short circuit transfer measurements made on the network are utilized and, in principle, these equal the actual short circuit admittances. The procedure for formulating equations becomes a straightforward and compact way of arriving at a complete set of equations. The number of equations obtained equals Q given in Eq. 1, even though in many cases the number of variables, i.e., B unknown elements of the network, will be less than Q. It was also shown in references that solution of the key subset of nonlinear equations derived from the M matrix implies solution of the network. This approach depends on a physical argument since it is assumed that the equations derived for the physical network using the specified partitioned matrix assure us of .consistency in the sense that there exist values of the unknown variables which satisfy all B_q equations. B_q is an upperbound for the number of equations. Thus

$$B_q = Q \geq B$$

Implications of the Key Subgraph and Solvability

The initial successes with the key subgraph concept for single-element-kind networks were found to be readily accessible to fault isolation in simple linear solid state amplifiers [D4]. These early efforts were primarily intended to determine theoretical limitations on solution techniques assuming idealized measurements and computations. Having found a means for systematically formulating the necessary set of key subgraph equations, it was shown that one could derive useful data for assumed measurement errors. An explicit example detailed in Bedrosian and Berkowitz [8] shows the twelve branch network for specified measurement errors of ± 5% for the key subgraph branches. Note in the resulting calculation, shown below, that the maximum deviation does not occur for the concealed

Transfer Measurements   Calculated Results

SC Admittance	Assumed Error		Network Branch	Ideal Case	Meas. Error	Deviation In %
Y_{13}	+5%					
Y_{15}	,,	B_k	a	1.000	0.959	- 4
Y_{25}	,,		b	1.000	0.959	- 4
Y_{26}	,,		d	2.000	2.064	+ 3
Y_{34}	,,		g	1.000	0.895	- 11
			h	2.000	2.357	+18
Y_{16}	-5 %		i	3.000	3.498	+17
Y_{24}	,,		j	1.000	1.271	+27
Y_{35}	,,		k	2.000	1.681	- 16
Y_{36}	,,		n	2.000	1.868	- 7
			x	6.000	6.021	+ 0
		B_f	y	4.000	4.013	+ 0
			z	4.000	3.920	- 2

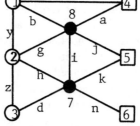

Figure 4. Network for detailed example and related computation.

branch joining the two internal nodes as may have been expected. This is a function of the magnitude of the errors in the specific measurements. One could readily extend this approach to extract useful reliability or design data.

The versatility of the key subgraph concept is further indicated by the fact that in general its solution implies solution of the family of solvable networks. This stems from the fact that the family of 2^{B_f} networks is related by having the same set of key subgraph equations. An example, using the twelve branch solvable network of Figure 4, is shown in Figure 5 [8].

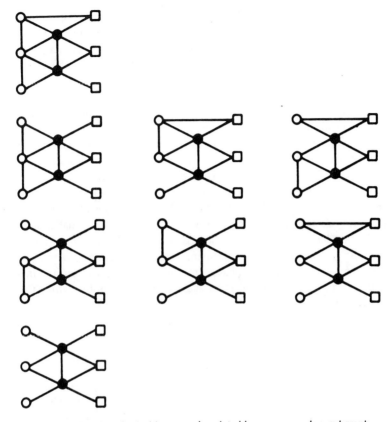

Figure 5. Family of solvable networks related by a common key subgraph.

As an aside, it is worth mentioning Darlington's [10] remark in 1955 concerning the theoretical problem of finding an entirely new method of treating networks with internal nodes. In particular the key subgraph concept greatly simplifies interpretation of Darlington's network equivalence theorem for RC networks. In fact a spin-off of the original work resulted in a simplified explicit solution of networks with two internal nodes [9]. The latter also impacts on the subsequent contribution by Even and Lempel [11]. They provide necessary and sufficient conditions for a node to be an essential terminal of the given network.

Even and Lempel also gave necessary and sufficient conditions for the required voltage measurements at a given node of a passive RLC network based on fault isolation with measurements made at a single frequency. Having derived some useful results with a carefully idealized model the logical extension is to introduce more realistic constraints. In any case, one must come to grips with the fact that one must live with "fuzzy" measurements and computations involving, at the minimum, round-off errors. For equipment designed with the more conventional lumped elements, one can, for example, attempt to take advantage of the fact that the electrical components are usually manufactured in standardized sets of values. Thus, various investigators introduced statistical considerations in connection with the fault isolation problem. It is clear that success in the effort would be important not only in fault isolation but also in giving a handle on failure prediction.

Having said that, we can indicate the relationship of maintainability via automatic test equipment to other important considerations in maintaining or improving operational effectiveness of an electronic system. A somewhat idealized capsule representation is given in Figure 6.

Another facet of the application of practical constraints is the fact that existing design procedures traditionally result in operational equipment with an inadequate number of accessible test points for a complete deterministic solution of the pertinent network equations. Attempts at overcoming such limitations have involved the application of inverse probability for fault isolation. Here again one is confronted with "fuzzy" results in that the

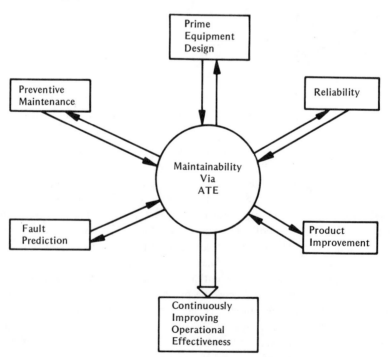

Figure 6. Interaction of maintainability with other facets of system activities.

techniques yield an indication of which component is most likely to be faulty rather than yielding a definite solution. Some investigators have attempted to deal with the fuzzy aspects of the problem as indicated above by resorting to heuristic methods.

A difficulty with many of the techniques that have been reported is that they are based in a crucial way on the assumption that only a single component has failed. A significant effort to overcome this limitation is the statistical estimation fault isolation procedure (SEFIP) which is designed to perform fault isolation by estimating values of all circuit parameters which are subsequently compared with "nominal" values [A5]. The techniques were shown to be workable for networks of limited size. The difficulty lies in the large amount of memory required even though the measurements and related computations can be made very quickly. This technique permits the use of simulation experiments.

Recently Saeks, Sing and Liu [12] introduced a general approach for simulating the components of such networks by augmenting the system under test with memoryless components. Ransom in his dissertation has extended their work [D15]. The result is a sufficient condition for fault isolation which turns out to be the equivalent of finding the left inverse of the connection function. It suffices for this discussion to consider the connection function as a convenient symbolic simplification of the usual system function used to describe the external behavior of a system. See the paper by Ransom in this Proceedings.

Alternative Approaches

Note the distinction between the solvability procedures described above and that proposed by Seshu and Waxman [6] wherein they compute the symbolic transfer function. Then to achieve fault detection and isolation nominal values of the circuit parameters are substituted into the symbolic transfer function to yield the poles and zeros of the function. Next each parameter is varied in discrete steps over a range of values and the magnitude of the transfer function is calculated for each test frequency. Programming details become important since thousands of calculations may be required. Their initial effort was based on the use of topological formulas involving tree and two-tree enumerations.

Subsequent to the untimely demise of Seshu, a progress report describing the pertinent large-scale computer program implementation was published by Maenpaa, et al [13]. This highlights the limitations of the use of topological formulas and tree generation techniques. Other techniques for obtaining symbolic transfer functions have since become available, such as the NASAP program and the SNAP program developed at Purdue.

Another alternative method for identifying multiple branch parameter changes was given by Neu [14]. Still a different approach has been advanced by Martens and Dyck [15]. They utilize the properties of the bilinear transformation for a network parameter for single fault identification using only magnitude and phase measurements along with precomputed loci plotted in the complex transfer function plane. We mention others without discussion [16-19].

Possible Extensions

We introduce this discussion by recalling some comments on the future of diagnosis made by the late Professor Seshu at a seminar more than a decade ago [A1]. With respect to theoretical developments, he felt there was a pressing need to develop procedures for generating the tests necessary to diagnose (sequential switching) circuits. Because this is closely connected to the problem of designing such circuits, he felt that we must get away from the very general and arbitrary circuits then in use. Seshu asserted that they are "hard to design, difficult to build, and impossible to diagnose." Hence, he predicted that there would be increased merging of diagnosis and design to the point that the designer would be required to supply the diagnostic tests for his circuits. Thinking in terms of ten to fifteen years ahead (which is where we are now) he felt it reasonable to automate the whole process so that the computer makes sure that the circuit being developed is diagnosable. He also stressed the need for theory to support diagnosis with imperfect test equipment. One might say that the theoretical models that he was envisioning fall under the general category of the pioneering work by Zadeh [20] dealing with "fuzzy" sets and related linguistic and other developments. A formal application of these concepts to the design of diagnostic procedures remains to be accomplished. In the realm of practical diagnostic systems Seshu gave as a conservative estimate the existence of some 60,000 small computers in the USA by 1974 (low by about a factor of two). This he felt would compel the computer manufacturers to seek an economical solution to the problem of mass maintenance. The several survey articles listed in the references indicate the intensity of such interest. Thus, it is clear that Seshu was not only a pioneer and significant contributor to this area, but also had tremendous insight into its future development. It is hoped that a spin-off of this symposium would help inspire some newcomers to this vital area to find the urgently needed workable solutions. Although most of the activity cited in the many references are from various organizations within the USA you should be aware that there is also intense interest in this area within other technologically advanced countries [21-25]. For example, the Soviet Union held the first "all union" conference on automated means of diagnosis in May 1973. Furthermore, both in the U.S.S.R. and in Japan one finds that engineers are investigating graph theoretic approaches to the difficult problems of fault isolation and fault analysis.

The availability of microprocessors provides greater flexibility in dealing with some aspects of the fault analysis problem. On the other hand we are now confronted with integrated circuits and LSI implementation of electronic systems. This suggests the need for in depth investigation of analytic procedures utilizing partially accessible nodes. There is also possible application for a dynamic interpretation to the topological domain chart representation [27]. This would permit us to take into account catastrophic changes such as short circuits and open circuits in the graph model of the analog circuits. Recall that the potential payoff can be in any of the facets of system operation indicated in Figure 6. A portion of the domain chart, for $B = M + N - 1$, is shown in Figure 7 including the cell locations for complete graphs denoted by K_N.

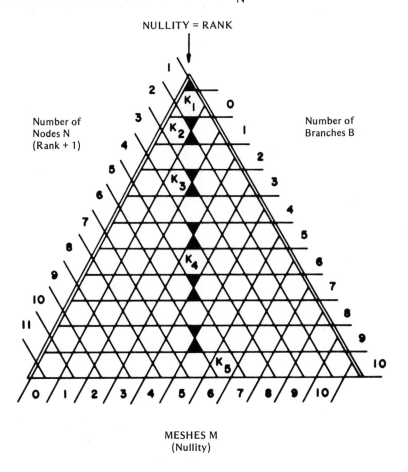

Figure 7. Symmetrical Domain Chart for connected linear graphs.

The above remarks are intended to convey some insight into the difficulties in dealing with fault analysis of analog circuits. It is very clear that the progress has been slow. As the growing number of survey papers indicates, the general topic of circuit checkout or fault diagnosis continues to be an area of growing importance. A key point is the fact that because of the unavoidable slow progress with fault analysis of analog circuits, we find ourselves in the transition phase of going from lumped discrete elements to integrated circuits and even LSI. Hence, we must report that the general problem remains unsolved and in fact technical advances have changed the character of the problem.

Notable progress has been made in dealing with diagnosis of digital circuits where graph theoretic approaches have also been utilized. Not surprisingly, it is also true that the use of large-scale integrated circuits is further along in the design of digital systems. There, also, one is concerned with developing design techniques to facilitate making the resulting circuits easily diagnosable. There is room, as well as growing motivation, to do further work in this area, both for treating multiple fault cases and for obtaining insight to provide design guidelines for automated diagnosable systems. Such systems can be considered forerunners to failure tolerant or self-repairing systems.

Because of the diversity and prolonged period during which there has been extensive effort at universities, industrial organizations, and the government, there has accumulated a vast literature in this area. In view of this we shall cite the references pertinent to this paper and also include some guidance as to other sources of information.

REFERENCES

1. American Tel. and Tel. Co., *Collected Papers of George A. Campbell*, New York, 1937, p. 78.

2. J. Riordan, "The Number of Impedances of an n-Terminal Network," Bell System Tech. Journal, Vol. 18, 1939, pp. 300-314.

3. B. McMillan, "Introduction to Formal Realizability Theory," Bell System Tech. Journal, Vol. 31, 1952, pp. 217 and 541.

4. IEEE Transactions on Computers, Special Issue on Failure Tolerant Computing, Vol. C-22, Mar. 1973.

5. C. Beckman, *et al*, "Study of Piece-Part Fault Isolation by Computer Logic," Final Report I.C.R., Univ. of Pa., Contract #DA36-304-507-ORD-3037RD, June 1960.

6. S. Seshu and R. Waxman, "Fault Isolation in Conventional Linear Systems-A Feasibility Study," IEEE Trans. on Realiability, Vol. R-15, May 1966, p. 11.

7. R. S. Berkowitz, "Conditions for Network-Element-Value Solvability," IRE Trans. Circuit Theory, Vol. CT-9, March 1962, pp. 24-29.

8. S. D. Bedrosian and R. S. Berkowitz, "Solution Procedure for Single-Element-Kind Networks," IRE International Convention Record, Vol. 10, Part 2, 1962, pp. 16-24.

9. S. D. Bedrosian, "A Simplified Explicit Solution of Networks with Two Internal Nodes," IEEE Trans. Communication and Electronics, Vol. 71, March 1964, pp. 219-224.

10. S. Darlington, "A Survey of Network Realization Techniques," IRE Trans. Circuit Theory, Vol. CT-2, Dec. 1955, pp. 291-296.

11. S. Even and A. Lempel, "On a Problem of Diagnosis," IEEE Trans. Circuit Theory, Vol. CT-14, Sept. 1967, pp. 361-364.

12. R. Saeks, S. P. Singh and R-W. Liu, "Fault Isolation via Components Simulation," IEEE Trans. on Circuit Theory, Vol. CT-19, No. 6, Nov. 1972, pp. 634-640.

13. J. H. Maenpaa, C. J. Stehman and W. J. Stahl, "Fault Isolation in Conventional Linear Systems: A Progress Report," IEEE Trans. on Reliability, Vol. R-18, No. 1, February 1969, pp. 12-14.

14. E. C. Neu, "A New n-Port Network Theorem," Proc. of the 13th Midwest Symp. on Circuit Theory, Minneapolis, 1970 (paper iv. 5).

15. G. O. Martens and J. D. Dyck, "Fault Identification in Electronic Circuits with the Aid of Bilinear Transformations," IEEE Trans. on Reliability, Vol. R-21, No. 2, May 1972, pp. 99-104.

16. J. E. D. Kirby, D. R. Towill and K. J. Baker, "Transfer Function Measurement Using Analog Modeling Techniques," IEEE Trans. on Instrumentation and Measurement, Vol. IM-22, No. 1, March 1973, pp. 52-61.

17. W. K. Chen and F. N. Chan, "On the Unique Solvability of Linear Active Networks," Trans. Circuits and Systems, Vol. CAS-21, Jan. 1974, pp. 26-35.

18. S. D. Shieu and S. P. Chan, "Topological Formulation of Symbolic Network Functions and Sensitivity Analysis of Active Networks," IEEE Trans. Circuits and Systems, Vol. CAS-21, Jan. 1974, pp. 39-45.

19. A. A. Wolf and J. H. Dietz, "A Statistical Theory for Parameter Identification in Physical Systems," J. of the Franklin Institute, Vol. 274, No. 1962, pp. 369-400.

20. L. A. Zadeh, "Toward a Theory of Fuzzy Sets," in *Aspects of Network and System Theory*, R. E. Kalman and N. DeClaris, Eds., Holt Rinehart and Winston, New York 1971, pp. 469-490.

21. T. Fukui and Y. Nakanishi, "A Method of Designing Output Sensors for Equipment Diagnosis Based on Information Content of Output Sensors," Electronics and Communications in Japan, Vol. 51-3, Jan. 1968, pp. 113-121.

22. E. S. Sogomonyan, "Monitoring Operability and Finding Failures in Functionally Connected Systems," translated from Avtomatika i. Telemekhankia, Vol. 25, No. 6, June 1964, pp. 980-990 (Moscow).

23. S. Hayashi, Y. Hattori and T. Sasaki, "Considerations on Network Element Value Evaluation," Electronics and Communications in Japan, Vol. 50, Dec. 1967, (transl. Oct. 1968), pp. 118-127.

24. S. A. Nechaev, "Length of Identification Test for a Type of Graphs," translated from Avtomatika i. Telemekhankia, No. 5, May 1973, pp. 172-176.

25. G. Dube and J-C. Rault, "La détection et la localisation des défauts dans les circuits analogiques," Revue Technique Thomson-CSF, Vol. 6, March 1974, pp. 71-80.

26. S. D. Bedrosian, "Applications of Graph Theory to Networks," in Proc. NATO Advanced Study Institute on *Network and Signal Theory*, J. Skwirzinski and S. Scanlan (eds.) Peter Peregriniun, U.K., 1973.

27. S. D. Bedrosian, "Application of Linear Graphs to Multi-level Maser Analysis," J. of the Franklin Institute, Vol. 274, No. 4, Oct. 1962, pp. 278-283.

A. Symposia, Proceedings and Project Reports

1. Battelle Memorial Institute, "Proceedings Seminar on Automatic Checkout Techniques," Sept. 1962. For a review of many of these papers by S. D. Bedrosian, see 2 below.

2. S. D. Bedrosian, "Paper Reviews on Automatic Checkout Techniques," IEEE Trans. on Electronic Computers, Vol. EC-13, April 1964, p. 164.

3. See also the Special Issue on Automatic Testing Techniques, IRE Trans. on Military Electronics, Vol. MIL-6, July 1962.

4. Proceedings of the International Conference on Aerospace Support, IEEE Transactions on Aerospace, Vol. AS-1, August 1963.

5. R. S. Berkowitz, W. G. Faust, and M. M. Vartanian, "Validation of Theoretical Automatic Checkout Techniques," Technical Report, AFAPL-TR-68-120, Oct. 1968, pp. 1-295.

6. J. Lustig and D. M. Goodman, "Trends in the Development of Automatic Test Equipment," Project SETE, Report 210/106, National Aeronautics and Space Administration, June 1973.

B. Survey Articles in the Trade and Professional Journals

1. R. Dobriner, "ACE: The Ultimate in Failure Detection," Electronic Design, Vol. 24, November 22, 1967.

2. J. Dent, "Diagnostic Engineering," IEEE Spectrum, Vol. 4, July 1967, p. 99.

3. S. E. Grossman, "Automatic Testing Pays Off. . . .", Electronics, Vol. 19, Sept. 1974, 95-109.

4. F. Liguori, Ed., *Automatic Test Equipment: Hardware, Software and Management*, IEEE Press, New York, to be published, Oct. 1974.

5. G. V. Novatny, "Automatic Checkout Equipment", Electronics, July 13, 1962, pp. 37-44.

6. M. Eleccion, "Automatic Testing: Quality Raiser, Dollar Saver," IEEE Spectrum, Aug. 1974, pp. 38-43.

7. K. To and R. E. Tulloss, "Automatic Test Systems," IEEE Spectrum, Sept. 1974, pp. 44-52.

8. G. Shapiro, *et al*, "Project FIST: Fault Isolation by Semiautomatic Techniques," Part 1, IEEE Spectrum, Vol. 1, Aug. 1964, p. 98 and Part 2, Sept. 1964, p. 130.

9. R. F. Garzia, "Fault Isolation Computer Methods," NASA Tech. Report CR-1758, George C. Marshall Space Flight Center, Huntsville, Alabama, 1971.

C. **Pertinent Short Courses**

1. Organized by D. M. Goodman of Project SETE at New York University, 1958, 1961 and 1968.

2. Special Summer Session, "Automatic Checkout Techniques," The Moore School of Electrical Engineering of the Univ. of Pennsylvania, June 1965.

3. "Computer Aided Testing and Fault Identification of Solid State Systems," Univ. of Wisconsin, May 1968.

D. **Some of the Pertinent Theses and Dissertations**

1. E. J. Kletsky, "Fundamental Limitations on Self-repairing Systems," Ph.D. Dissertation, Syracuse University, 1961.

2. S. D. Bedrosian, "On Element Value Solution of Single-Element-Kind Networks," Ph.D. Dissertation, University of Pennsylvania, Dec. 1961.

3. L. J. Buchsbaum, "Fault Diagnosis by Computational Methods," MSE Thesis, The Moore School of Electrical Engineering of the Univ. of Pennsylvania,Dec. 1963.

4. R. L. Gayer, "Fault Isolation in Solid State Circuits," MSE Thesis, The Moore School of Electrical Engineering of the Univ. of Pennsylvania, Aug. 1963.

5. M. M. Vartanian, "Validation Study of an Electronic Network Fault Isolation Technique to Automatic Control Systems," MSE Thesis, The Moore School of Electrical Engineering of the Univ. of Pennsylvania, Dec. 1967.

6. S. M. Worthington, "Application of Automatic Checkout and Fault Isolation Techniques to Automatic Control Systems," MSE Thesis, The Moore School of Electrical Engineering of the Univ. of Pennsylvania, Dec. 1967.

7. S. P. Schechnea, "Management Strategy and Policy in Fault Isolation Techniques," MBA Thesis, Wharton School of the Univ. of Pennsylvania, 1968.

8. S. Srivastava, "Development of Fault Isolation Procedures for Semi-automatic Fault Isolation of Electronic Assemblies," MSE Thesis, The Moore School of Electrical Engineering of the Univ. of Pennsylvania, Aug. 1968.

9. W. G. Faust, "The Application and Development of Search Techniques for Fault Isolation," Ph.D. Dissertation, University of Pennsylvania, May 1969.

10. P. H. Jackson, "Fault Isolation by Transfer Characteristics," MSE Thesis, The Moore School of Electrical Engineering of the Univ. of Pennsylvania, May 1969.

11. E. Levinson, "A Direct Method of Fault Isolation," MSE Thesis, The Moore School of Electrical Engineering of the Univ. of Pennsylvania, May 1969.

12. K. Loewenstein, "Design of Automatic Functional Tester for a Miniaturized AM Radio Integrated Circuit," MSE Thesis, The Moore School of Electrical Engineering, University of Pennsylvania, Aug. 1969.

13. M. M. Vartanian, "An Algorithm for Fault Isolation of Multi State Electronic Networks," Ph.D. Dissertation, University of Pennsylvania, 1969.

14. M. R. Garey, "Optimal Binary Decision Trees for Diagnostic Identification Problems," Ph.D. Dissertation, University of Wisconsin, 1970.

15. M. N. Ransom, "A Functional Approach to the Connections of a Large Scale Dynamical System," Ph.D. Dissertation, Notre Dame University, May 1973.

REAL-TIME FAULT DETECTION AND FAULT-TOLERANT
IMPLEMENTATIONS FOR SEQUENTIAL CIRCUITS

Gary K. Maki
Dwight H. Sawin III

Introduction

Real-time fault detection and fault tolerance has become increasingly more important to the digital circuits designer in recent years. This paper is concerned with the fault detection and tolerant design aspects for synchronous and asynchronous sequential circuits. Much of the material presented in this paper has been proven elsewhere and therefore an overview and a discussion concerning the means of designing fault detecting and fault tolerant circuits are discussed.

In attempting to implement a fault-detecting or fault-tolerant sequential circuit, there are several important concerns the digital circuit designer should address in selecting a design technique.

●The range of fault conditions covered should be large. Not only should the resulting design be capable of providing protection for the classical stuck-at-faults, but should include bridging faults, shorts, open circuits, intermittent faults, and failures due to noise.

●The additional circuitry required to achieve a specified level of protection must not be excessive. If the hardware requirements are excessive, overall reliability can be less than that of a regularly designed circuit. Construction of two identical circuits would provide fault detection, and use of three circuits together with a majority voter (triple modular redundancy, TMR) provides single fault tolerance. Therefore, these simple approaches can serve as upper bounds on the hardware requirements as far as a practical implementation is concerned.

Gary K. Maki is with the Electrical Engineering Department, University of Idaho, Moscow, Idaho 83843.

Dwight H. Sawin III is with the Communications/ADP Laboratory, USAECOM, Ft. Monmouth, N. J. 07703.

This research was supported in part by the Office of Naval Research under contract N00014-72-A-0232-0002 NR 048-617.

●The design procedure should be relatively simple and be applicable to the types of circuit problems usually encountered. Furthermore, the design procedure would be more convenient if it is not necessary to require the use of computers.

●The resulting circuit should be void of lurking faults [1]. A lurking fault does not affect normal circuit operation in any way when it occurs to the redundant circuitry which provides protection against faults. When a lurking fault is present, a second fault can cause a catastrophic failure when the circuit is forced to utilize the protective circuitry which has failed previously. The protective circuitry should be self-checking or checked by a separate means.

●Special hardware components should be avoided. Some applications require that the components used must meet rigid specifications and that extensive testing must be carried out to certify the use of any special components. This process can be costly and time consuming.

Sequential Circuits

Asynchronous sequential circuits are normally defined in terms of a flow table such as shown in Fig. 1. The circuits here are assumed to operate in the normal fundamental mode and are encoded with single-transition-time (STT) state assignments [2]. These are the practical constraints usually placed on these circuits and allow for fastest possible circuit operation.

A few definitions and terminology are in order to allow for a presentation of the design algorithm that follows. Shown in Fig.1 is a STT assignment for Machine A. Each state variable y_i can be defined in terms of a τ-partition, τ_i, which partitions those states where $y_i = 1$ in the first block from those states where $y_i = 0$ in the second block. For example, $\tau_1 = \langle BDE; ACF \rangle$. The two-block η-partition, η_i^p, partitions those states where the next state variable Y_i is 1 in the first block from those where $Y_i = 0$ in the second block under input I_p. For example, $\eta_1^1 = \langle EF; ABCD \rangle$ (Y_1 is 1 in states E and F, and 0 in A,B,C, and D under I_1). The two-block partition, γ_j^p, partitions those states where $Z_j = 1$ from those where $Z_j = 0$ under I_p. For example, $\gamma_1^1 = \langle ABCD; EF \rangle$ (the unstable state is considered to have the same output as the stable state).

The design equations are often expressed as shown in Eq. (1),

$$Y_i = f_{i1}(y)I_1 + f_{i2}(y)I_2 + \ldots + f_{im}(y)I_m$$

$$Z_j = g_{j1}(y)I_1 + f_{j2}(y)I_2 + \ldots + g_{jm}(y)I_m$$

Eq. (1)

where Y_i is the next stable variable, $f_{ip}(y)$ and $g_{jp}(y)$ are functions of the state variables only, I_p denotes the input state, and Z_j the output. A *k-set* is a collection of the k states that have the same next state entry under an input. For example, the k-sets under I_1 are AB, CD, and EF.

The class of synchronous sequential circuits considered here is level input-level output circuits with synchronizing clock pulses. Figure 10 is an example of a flow table showing this type of circuit.

y_1	y_2	y_3		I_1	I_2	I_3
0	0	0	A	Ⓐ/1	F	D
1	0	0	B	A	Ⓑ/0	D
0	1	1	C	Ⓒ/1	F	Ⓒ/1
1	0	1	D	C	B	Ⓓ/0
1	1	0	E	Ⓔ/0	B	C
0	1	0	F	E	Ⓕ/1	C

Figure 1. Asynchronous Sequential Circuit-Machine A

Fault Detection

The problem of real-time fault detection in asynchronous sequential circuits has been solved a number of different ways [3,4,5]. A necessary design constraint for the detection of a single fault in asynchronous sequential circuits is that the minimum distance between transition paths be at least 2. A first approach to realizing a fault detecting design is to find an STT assignment where the minimum distance is 2. Shown in Fig. 2 is a single column of a flow table with a satisfactory state assignment that partitions the transition paths a distance of 2.

y_1	y_2	y_3	y_4	y_5		I_p
0	1	0	1	0	A	Ⓐ
1	1	1	1	0	B	A
1	1	0	1	1	C	A
0	1	1	0	1	D	Ⓓ
1	0	1	0	1	E	D

Figure 2. Fault Detection Example

Two effects occur when a fault is present. First the circuit remains in a transition path and becomes stable, but not in the proper stable state. The second effect forces the circuit out of the transition path. With the transition paths a distance 2, ⁵detection of the second type of fault necessitates assigning the non-transition path states to be stable. Therefore, when the circuit becomes stable, a check can be made whether or not the stable state is the proper stable state; if not in the proper stable state, a fault has occurred. The state table assigned to accomplish this task is shown in Fig. 3 with the transition paths denoted (●). The next state equations for this input are

$$Y_1 = y_1 y_4' y_5' + y_1 y_3' y_4' + y_1 y_2' y_4 + y_1 y_3 y_4 y_5$$
$$Y_2 = y_2 + y_3 y_4' y_5$$
$$Y_3 = y_2' y_3 + y_3 y_4' + y_3 y_5$$
$$Y_4 = y_4$$
$$Y_5 = y_3 y_5 + y_4' y_5$$

The fault detector would then be a combinational logic circuit that denotes whenever the circuit is not in a proper stable state. This is shown in Figure 4, and the expression for D is

$$D = y_3' y_4' + y_4 y_5 + y_1 + y_2' + y_3 y_5'$$

There are several disadvantages with this approach from the implementation viewpoint. One is the relative difficulty inherent in arriving at the design equations. It is necessary to use a state table, however, even working with small flow tables can produce an unmanageable number of state variables on a K-map.

Another limitation is the amount of logic required. Use of this approach will easily exceed twice the amount of logic required to realize a

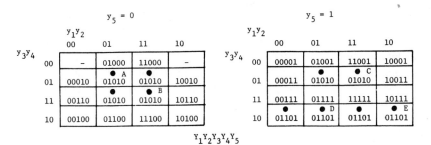

Figure 3. State Table of Figure 2

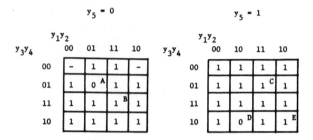

Figure 4. Fault Detector of Figure 2

Figure 5. Lurking Fault $y_3 y_5$ s-a-0

single copy of the circuit, even ignoring the detector. In addition to this, there exist lurking faults which are undetectable during normal operation. For example, if the term $y_3 y_5$ in Y_3 became s-a-0, only states 01111 and 11111 are affected by the fault. As shown in Fig. 5, these are non-transition path states. Therefore, circuit operation would proceed normally with this fault condition present. In general, a lurking fault could be defined as one which affects only non-transition path states. In this example there are numerous lurking faults. Any two of the three terms in Y_3 can be s-a-0 and the entire (or any part of the) expression for Y_1 can be s-a-0 and be undetectable. Therefore, from the moment power is applied to the circuit, it is impossible to detect the presence of these fault conditions. Furthermore, one could not guarantee that a later fault condition would be detected. The only way to check for lurking faults would be to halt circuit operation and apply input test sequences which would force the circuit out of the normal set of states (special circuitry would have to be added to accomplish this), thereby defeating the objective of providing for real-time fault detection.

    With a lurking fault present, the circuit could malfunction without giving any prior indication that anything was wrong. For example, let $y_3 y_5$ of Y_3 be s-a-0, a lurking fault. Then while the circuit is in state D, let y_4

become s-a-1. The circuit will first transition to state 01111 and then proceed, after two gate delays in a NAND-NAND gate realization, to state 01011 at which time it is excited to state A 01010. A race between the logic that makes up the detector D and state variable y_5 exists, and if y_5 changes faster, then D will not indicate that the circuit has transitioned from state D to A improperly.

Ignoring lurking faults, further problems exist in the detection of a fault. It is necessary to wait for the circuit to settle down as it goes from an unstable to a stable state before a fault condition is known. This means an extra delay is required before determining the status of D. Secondly, the logic that makes up the detector must be hardcore.

The following is another procedure for achieving real-time fault detection in asynchronous sequential circuits that does not suffer from the above disadvantages [5,6]. The intent is not to go through the associated theory but to present the design algorithm and point out applicability to practical problems. The principal advantages of this design are

1. A large range of fault conditions is covered.
2. The design equations are obtained by inspection.
3. There are no lurking faults.
4. The fault detector has little hardcore.
5. Fault conditions are known after two gate delays.
6. Excessive logic is not required.

The design algorithm follows:

1. List the k-sets for each input state.
2. With x k-sets for an input I_p, form the τ-partitions (state variables) for each input using Eq. (2).

$$\tau_1 = \langle k_1 ; k_2, k_3, \ldots k_x \rangle$$

$$\tau_2 = \langle k_2 ; k_1, k_3, \ldots k_x \rangle$$

$$\bullet$$
$$\bullet \qquad\qquad (2)$$
$$\bullet$$

$$\tau_x = \langle k_x ; k_1, k_2, \ldots k_{x-1} \rangle.$$

3. Form the η_i^p partitions for each input I_p.

4.   Find the design equation coefficients f_{ip} [Eq. (1)] using the following:

    (A)   If all the elements of η_i^p are in one block, then f_{ip} = 1 or 0.

    (B)   Let τ_i, i = 1,2, . . . ,x partition the x k-sets k_i of I_p in the form of Eq. (2). For all f_{ip} terms, i = 1,2, . . . ,x; f_{ip} = y_i. For j $\neq$ $\langle 1,2, . . . ,x \rangle$, if η_j^p = τ_i, then f_{jp} = y_i; if η_j^p = $\langle k_1, k_2, . . . ,k_r;$ $k_{r+1}, . . ,k_x \rangle$, r $<$ x, then f_{jp} = $y_1 + y_2 + . . . + y_r$.

5.   The output is found by

    (A)   Form output partition γ_i^p .

    (B)   If γ_i^p = $\langle k_1, k_2, . . . , k_r; k_{r+1}, . . . ,k_x \rangle$, r$<$x, then g_{ip} = $y_1 + y_2 + . . . + y_r$.

Even though this procedure may appear cumbersome at first glance, it is easy to use and is as efficient for large problems as small ones. Consider the example of Fig. 1 to illustrate this procedure.

The k-sets under each input are

I_1	I_2	I_3
k_1 = AB	k_4 = ACF	k_6 = ABD
k_2 = CD	k_5 = BDE	k_7 = CEF
k_3 = EF		

The τ-partitions generated from these k-sets are

I_1	I_2	I_3
τ_1 = $\langle$AB; CDEF$\rangle$	τ_4 = $\langle$ACF; BDE$\rangle$	τ_6 = $\langle$ABD; CEF$\rangle$
τ_2 = $\langle$CD; ABEF$\rangle$	τ_5 = $\langle$BDE; ACF$\rangle$	τ_7 = $\langle$CEF; ABD$\rangle$
τ_3 = $\langle$EF; ABCD$\rangle$		

The partitioning variables for I_1, I_2, and I_3 are $\langle \tau_1, \tau_2, \tau_3 \rangle$, $\langle \tau_4, \tau_5 \rangle$, and $\langle \tau_6, \tau_7 \rangle$ respectively. $\langle \tau_1, \tau_2, \tau_3 \rangle$, $\langle \tau_4, \tau_5 \rangle$, and $\langle \tau_6, \tau_7 \rangle$ are known as the partition sets of I_1, I_2, and I_3 respectively. The η-partitions for each next state variable are

	I_1	I_2	I_3
η_1	AB; CDEF	BDE; ACF	ϕ; ABCDEF
η_2	CD; ABEF	ϕ; ABCDEF	ABCDEF; ϕ
η_3	EF; ABCD	ACF; BDE	ϕ; ABCDEF
η_4	ABCD; EF	ACF; BDE	CEF; ABD
η_5	EF; ABCD	BDE; ACF	ABD; CEF
η_6	AB; CDEF	BDE; ACF	ABD; CEF
η_7	CDEF; AB	ACF; BDE	CEF; ABD

The next state equations are

$$Y_1 = y_1 I_1 + y_5 I_2$$
$$Y_2 = y_2 I_1 + I_3$$
$$Y_3 = y_3 I_1 + y_4 I_2$$
$$Y_4 = (y_1 + y_2)I_1 + y_4 I_2 + y_7 I_3$$
$$Y_5 = y_3 I_1 + y_5 I_2 + y_6 I_3$$
$$Y_6 = y_1 I_1 + y_5 I_2 + y_6 I_3$$
$$Y_7 = (y_2 + y_3)I_1 + y_4 I_2 + y_7 I_3$$

These equations are obtained from step 4. A few words of illustration may help. First, $\eta_2^2 = \eta_1^3 = \eta_3^3 = \langle \phi;\ ABCDEF \rangle$; therefore $f_{22} = f_{13} = f_{33} = 0$. Since $\eta_2^3 = \langle ABCEDF;\ \phi \rangle$, $f_{23} = 1$. For the input I_p where y_i is in the partition set of I_p, $f_{ip} = y_i$. This is seen for the partition set of I_1 $\langle y_1, y_2, y_3 \rangle$ where $f_{i1} = y_i$, $i = 1,2,3$; the same is true for the partition sets of I_2 and I_3, $\langle y_4, y_5 \rangle$ and $\langle y_6, y_7 \rangle$ respectively. It has been shown [5] that for each partitioning variable of an input I_p, $f_{ip} = y_i$. Therefore, since each partitioning variable depends only on itself, a single fault can affect no more than one partitioning variable.

Variables y_4, y_5, y_6, and y_7 are non-partitioning variables in I_1. Since $\eta_4^1 = \langle ABCD;\ EF \rangle\ (\langle k_1, k_2,\ k_3 \rangle)$, by 4B, $f_{41} = y_1 + y_2$. Likewise, $\eta_5^1 = \langle EF; AB$ $CD \rangle$, $(\langle k_3;\ k_1, k_2 \rangle)$, $f_{51} = y_3$. The other non-partitioning variables are obtained in a like manner.

The output is obtained by first listing the γ_i^p partitions.

	I_1	I_2	I_3
γ	ABCD; EF	ACF; BDE	CEF; ABD

The output then is

$$Z = (y_1 + y_2)l_1 + y_4 l_2 + y_7 l_3$$

This expression is obtained the same way the non-partitioning variables are. For example, $\gamma_1^1 = \langle ABCD; EF \rangle (\langle k_1, k_2 ; k_3 \rangle)$, then $g_{11} = y_1 + y_2$.

y_3y_4 \ y_1y_2	00	01	11	10
00	–	01011	11010	–
01	00010	●A 01010	● 01010	10010
11	00110	● 01010	●B 01010	10110
10	00110	01110	11110	10110

y_3y_4 \ y_1y_2	00	01	11	10
00	00011	01011	11011	10011
01	00011	● 01010	● 01010	10011
11	00111	01011	11011	10111
10	● 01111	●D 01111	● 01111	●E 01111

Figure 6. Circuit Malfunction with Lurking Fault Present

A NAND gate realization of Machine A is shown in Fig. 7. Sharing logic is permissible and it is this aspect that yields an advantage of this procedure relative to hardware requirements. It has been shown that the upper bound for hardware requirements using this technique is the same as that of a regularly designed circuit without fault detection considerations [5].

Several points have been shown concerning the nature of the design equations [5,6].

(1)   The parity of each partition set is odd under fault free conditions.

(2)   For any single fault, at most only one partitioning variable is affected.

(3)   Whenever a fault affects a member of the partition set of l_p, it will affect no more than one elements of partition sets of other inputs.

(4)   The output is a function only of the partitioning variables.

From (1), whenever the parity of the partition set of l_p is even when the input is l_p, a fault condition is present. Furthermore from (2), only one variable of a partition set is affected by a fault. Therefore, a fault detector of the form shown in Fig. 8 can be implemented.

Effective fault detection is achieved by examining the parity of the partition set of the input that the circuit is in. For example, when the input is l_1, variables y_1, y_2, and y_3 must display odd parity (100, 010, or 001) for fault free conditions. An important aspect of these circuits is that whenever the input is l_i, the variables of the partition set of l_i are not excited and do not change, even when the circuit is going from unstable to stable state. The practical consequence of this is that one does not have to wait for the circuit to settle down (i.e., assume a stable state) before detecting the presence of faults. In fact, detection can be achieved in two gate delays; the AND and OR gate of Fig. 8.

A good question concerns the detection of a fault that affects only non-partitioning variables. First, since non-partitioning variables do not feed back, a single fault can affect only one non-partitioning variable. Secondly, from (4) above, the output is not affected. If y_i of the partition set of I_q is the non-partitioning variable that is affected by the fault, then if the input changes to I_q, the fault will be detected since the partition set of I_q is even. On the other hand, if the next input is other than I_q, then the fault has no effect on circuit operation and is tolerated. Only when the input is I_q will the fault affect circuit operation and only then does its presence need be known.

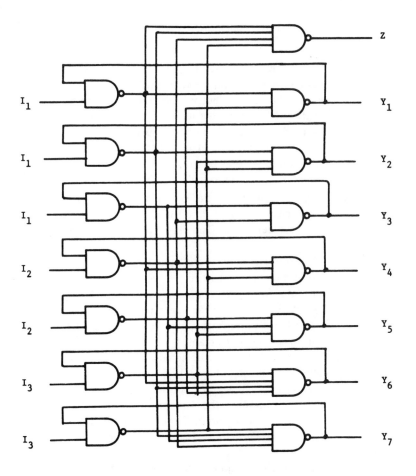

Figure 7. NAND Gate Realization of Machine A

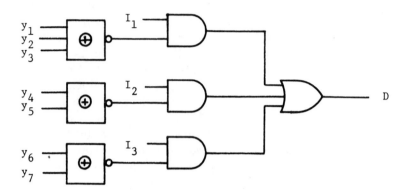

Figure 8. Fault Detector of Machine A

The only critical fault in Fig. 8 is the output D s-a-0. All other component failures are tolerated and internal fault conditions are detectable. For example, assume that the output of the complemented exclusive OR gate, or output of the AND gate or the input of the OR gate of the circuitry that examines the partition set if I_1 is s-a-0. Furthermore, let y_1 be s-a-1 while the circuit is in state E of I_1. The state that the circuit assumes is 1011111, where the correct state is 0010101. When the input changes to I_2 or I_3, $y_4 y_5 = 11$ or $y_6 y_7 = 11$ and the fault is detected.

In general, from (3), since a fault affects no more than one element of a partition set, any one element of the detector can fail and internal faults of the sequential circuit are detectable. The only exception is the output being s-a-0.

Lurking faults do not exist since the same logic that provides for fault detection causes the circuit to sequence through the states. Both circuit operation and fault detection are determined by the value of the partitioning variables. Since the partitioning variables are a function of only themselves $(f_{ip} = y_i)$, it is impossible for a fault to affect a partitioning variable and not affect circuit operation and be undetectable.

Faults in the output circuitry are detected by duplication of the output gate. This duplication is depicted in Fig. 9. If $Z = y_i I_p + \cdots$, then two copies are realized as shown in Fig. 9. The only way both copies are affected by a fault is if variable y_i fails; however, the detector of Fig. 8 would detect this fault condition. Both copies of an output Z_1 and Z_2 are fed into an exclusive OR gate and inputed to the OR gate of Fig. 8; this part of the total detector is hardcore, however.

In the discussion above, static hazards were not considered. Static hazards [2] can easily create problems in asynchronous sequential circuit designs. Static hazards exist only during inter-column transitions and not between transitions from unstable to stable state. Static hazards are easily eliminated. Let the next state equation for Y_i be of the form

$$Y_i = y_i I_p + y_j I_q + y_k I_r + y_\varrho I_s + \cdots$$

where y_i is a partitioning variable of I_p and y_j, y_k and y_ϱ are partitioning variables of $I_q, I_r,$ and I_s. Assume there are input transitions between I_q to I_p and I_r to I_p, but not I_s to I_p. Static hazards are eliminated by adding the following terms to the expression for Y_i:

$$y_i y_j + y_i y_k.$$

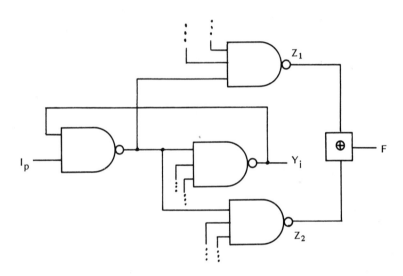

Figure 9. Output Detector $Z = y_i I_p + \cdots$

The detector depicted in Fig. 8 does not in general detect the presence of a failure in a hazard eliminating term. However, if a hazard eliminating term does fail and the subsequent hazard forces a partitioning variable to assume an incorrect value, then the detector will detect the malfunction. All hazard malfunctions — dynamic, essential, and function — are detectable.

Only a slight modification is needed to achieve a fail-safe asynchronous sequential circuit [7]. A fail-safe circuit is designed such that at the occurrence of a fault, only the output state (1 or 0) which results in the least loss can occur. An asynchronous sequential circuit is a 1(0)-fail-safe circuit if any failure to either the internal state or output logic causes only incorrect

1(0) and correct 0(1) outputs. The key to achieving a fail-safe design is obtaining an independent copy of the output. A second output is achieved as follows: Consider the set of partitioning variables y_i, $i = 1,2, \ldots ,x$, of Eq. (2). For each $\eta_i^p = \langle k_1, k_2, \ldots ,k_r; k_{r+1}, \ldots ,k_x \rangle$, g_{ip} can be expressed two ways: $g_{ip} = y_1 + y_2 + \cdots + y_r$ and $g_{ip} = y'_{r+1} \cdots y'_x$. Using this concept, the second copy of the output for machine A is

$$Z = y'_3 l_1 + y'_5 l_2 + y'_6 l_3$$

The second copy is a function of a different set of partitioning variables than the first copy. It has been shown [7] that a single fault cannot affect both copies of the output, even after many input changes.

A 1-fail-safe circuit is obtained by ORing the two copies of the output together. If one copy assumes a false 0, the correct copy will produce the correct output. If a false 1 is present on one copy, the output would assume the false 1 value; hence 1-fail-safe. A 0-fail-safe circuit is achieved by ANDing the two copies of the output together. The use of dotted logic for the output OR gate or AND gate will further improve the reliability.

			X		
$y_1\ y_2\ y_3$		0	1	Z
0 0 0	A	A	D	1
1 1 0	B	C	D	0
0 1 1	C	D	A	0
1 0 1	D	A	B	0

Figure 10. Synchronous Sequential Circuit-Machine C

Synchronous Sequential Circuits

Any minimum distance 2, code state assignment is sufficient to produce fault detection. Shown in Fig. 10 is a synchronous table with a minimum distance 2 assignment. If clocked D flip-flops are used, the D inputs and Z would be

$$D_1 = y'_1 y_2 x' + (y_1 + y'_3)x$$
$$D_2 = y_1 y'_3 x' + y_1 y'_2 x$$
$$D_3 = y_2 x' + y'_3 x$$
$$Z = y'_1 y'_2$$

In this paper the 1-out-of-n code will be used for fault detection. For Machine C, Fig. 10, the assignment is

y_1	y_2	y_3	y_4	
1	0	0	0	A
0	1	0	0	B
0	0	1	0	C
0	0	0	1	D

More flip-flops are required to realize this assignment, but there are other advantages: First the state assignment and design equations are obtained by inspection making this procedure attractive in working large problems. Furthermore, it is relatively efficient in hardware. For instance, the D input and output equations are

$$D_1 = (y_1 + y_4)x' + y_3 x$$
$$D_2 = y_4 x$$
$$D_3 = y_2 x'$$
$$D_4 = y_3 x' + (y_1 + y_2)x$$
$$Z = y_1$$

Only 20 gate inputs are required here whereas 25 are needed with the first procedure.

The detection of a fault condition is accomplished by observing a change in parity on the state variables. To guarantee that a single fault will affect no more than one state variable, it is not permissible to share logic between the state variables. Detection of faults on the output requires duplication of the output logic.

By examination of the design equations, it should be clear that lurking faults do not exist. If any state variable, or any element that makes up the D input assumes a false value, the circuit will assume an improper state and be detected.

Fault Tolerance

A circuit is single fault-tolerant if correct outputs are derived even in the presence of a single fault. There are several desirable features one looks for in single fault-tolerant designs. Besides those features stated earlier, it is desirable to be able to detect tolerated faults. It is important to know when a single fault-tolerant circuit has suffered a single fault in order that repairs can be initiated. Furthermore, it is helpful to know when the fault-tolerant capability has been exceeded; in worst case situations, two faults exceed the capability of a single fault-tolerant circuit.

Asynchronous Sequential Circuits

Single fault-tolerance in sequential circuits has been achieved with minimum distance 3 state assignment [4]. This procedure is illustrated with Machine D of Fig. 11. In Fig. 12 is shown the state table for $X = 0$. The next state entries of the transition paths are assigned in the normal fashion. The next state entries for all states adjacent to the transition paths are assigned the same as the transition paths. For example, the 8 states in the transition path for AB, $T(AB)$, are 101 (means all combinations of 1's and 0's) and have the next state entry of state A 101101. The states 100 are adjacent to $T(AB)$ and are also assigned the next state entry 101101. This procedure allows the circuit to arrive at the proper stable state (or at worst, a distance 1 from it) even if one variable assumes a false value.

The next state equations for this design under $X = 0$ are

$$Y_1 = y_1 y_2' x' + y_2' y_3 x' + y_1 y_3 x'$$

$$Y_2 = y_1' y_2 x' + y_2 y_3' x' + y_1' y_3' x'$$

$$Y_3 = y_1 y_2' x' + y_2' y_3 x' + y_1 y_3 x'$$

$$Y_4 = y_1 y_2' x' + y_2' y_3 x' + y_1 y_3 x'$$

$$Y_5 = y_1' y_2 x' + y_2 y_3' x' + y_1' y_3' x'$$

$$Y_6 = y_1 y_2' x' + y_2' y_3 x' + y' y_3 x'$$

y_1 y_2 y_3 y_4 y_5 y_6							X 0	1
1	0	1	1	0	1	A	(A)	D
1	0	1	0	1	0	B	A	(B)
0	1	0	0	1	0	C	(C)	B
0	1	0	1	0	1	D	C	(D)

Figure 11. Machine D

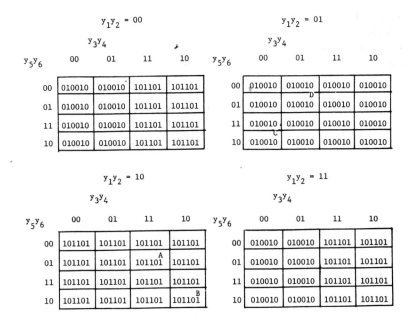

Figure 12. State Table for Machine D

Even though next state variables are equal under this input, it is not permissible to share logic. If shared logic was used, a single fault could force the circuit more than a distance 1 from a transition path and not be tolerated. Not allowing shared logic makes the hardware requirements excessive.

As with the case of fault detection, lurking faults exist. If any two of the three terms in each of the next state equations for Y_1, Y_2, Y_3, Y_4, Y_5, and Y_6 becomes s-a-0, it would be impossible to detect the fault condition since the circuit would not enter the affected states. The situation could occur in which all the added protective circuitry can fail without any observable change in circuit behavior. To illustrate this point, let the input y_3' in the term $y_1' y_3' x'$ in Y_2 be s-a-1, a lurking fault. The result is $Y_2 = y_1' x' + y_2 y_3' x'$. Now if Y_1 became s-a-o while the circuit was in state A, the circuit will race to state D and become stable. This is an example of not being able to determine the presence of a fault condition in a single fault-tolerant circuit and the circuit malfunctioning at the occurrence of another fault. For all practical purposes, it is impossible to know when the first fault occurs in this single fault-tolerant design.

The approach presented now, though conceptionally not very elegant, is very effective. The concept is depicted in Fig. 13 in which two independent asynchronous sequential circuits, each capable of detecting fault conditions, are used. This scheme is effective if fault detection is easily and efficiently achieved and if the switch is relatively simple. Fault detection is easily achieved as pointed out earlier and the switch can be a simple SR flip-flop.

The best-bound relative to the number of gate inputs needed to realize an asynchronous sequential circuit [2] is given as

$$\Sigma h_n + d_s(m+1) + mt \qquad 1 \leqslant n \leqslant d_s \qquad (3)$$

where

d_s = number of distinct non-trivial k-sets
h_n = number of stable states that are contained in the nth k-set
m = number of input variables
t = number of trivial input columns

The bound for a fault detecting circuit, including detectors is

$$\Sigma h_n + d_s(m+2) + mt \quad 1 \leq n \leq d_s \qquad (4)$$

Three times the value of Eq. (3) (plus a few gate inputs for the voter) is the best bound with TMR. The bound for the procedure illustrated here is twice

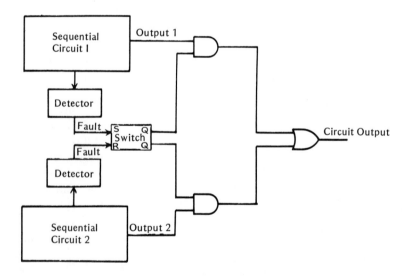

Figure 13. Single Fault Tolerant Design

the value of Eq. (4) (plus a few gate inputs for the flip-flop and associated gates). Clearly the bound for the circuits presented here is less than that of TMR. Furthermore, it is clear that fault conditions are detectable and one knows when the fault capability is exceeded.

Synchronous Sequential Circuits

Design techniques for realizing single fault-tolerant synchronous sequential circuits have been developed [8,9]. A minimum distance 3 state assignment is used, and with proper assignment of the state table the design equations are formed. The state assignment for a single fault-tolerant design of Machine C is shown below.

y_1	y_2	y_3	y_4	y_5	
0	0	0	0	0	A
1	1	0	1	0	B
0	0	1	1	1	C
1	1	1	0	1	D

Shown in Fig. 14 is the properly assigned state table where each state of the flow table and each adjacent state are assigned the same next state entry. This procedure will allow correct circuit operation even if one state variable

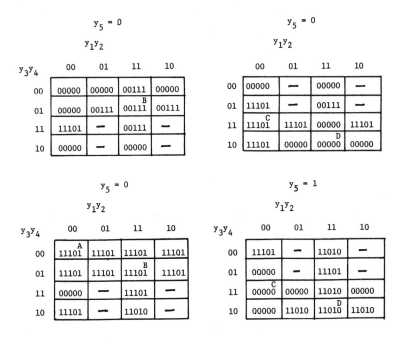

Figure 14. Fault Tolerant State Table of Machine C

assumes a false value. For example, the next state entry for 11010 (B), 01010, 10010, 11110, 11000, and 11011 is 00111, the code for C. Should a single fault occur, the circuit would assume a state no greater than a distance 1 from the state under fault free conditions, and since all these states have the same next state entry, a single fault would be tolerated. The design equations for the state table of Fig. 14 are

$$Y_1 = (y_1' y_4 y_5 + y_1' y_2' y_3 y_5 + y_2' y_3 y_4)x' + (y_3' y_4' + y_1 y_2 + y_1 y_4' + y_2 y_4' + y_4' y_5' + y_3' y_5')x$$

$$Y_2 = (y_1' y_4 y_5 + y_1' y_2' y_3 y_5 + y_2' y_3 y_4)x' + (y_3' y_4' + y_1 y_2 + y_1 y_4' + y_2 y_4' + y_4' y_5' + y_3' y_5')x$$

$$Y_3 = y_1 y_2 y_3' y_5' + (y_1' y_3 y_4 + y_2 y_4 y_5' + y_1 y_2' y_4 + y_1' y_2' y_3 y_5 + y_3' y_4 y_5)x' + (y_3' y_4' + y_1 y_2 + y_1 y_4' + y_2 y_4' + y_4' y_5' + y_3' y_5')x$$

$$Y_4 = (y_1 y_4 y_5' + y_2 y_4 y_5' + y_1 y_3' y_4 + y_1 y_2 y_3' y_5')x' + (y_1 y_3 y_4' + y_2 y_3 y_4' + y_1 y_2 y_3 y_5 + y_1 y_4' y_5)x$$

$$Y_5 = y_1 y_2 y_3' y_5' + (y_1' y_3 y_4 + y_2 y_4 y_5' + y_1 y_2' y_4 + y_1' y_2' y_3 y_5 + y_3' y_4 y_5)x' + (y_3' y_4' + y_1 y_2 + y_1 y_4' + y_2 y_4' + y_4' y_5' + y_3' y_5')x$$

It is not permissible to share logic; if logic was shared, a single fault could allow the circuit to assume a state a distance greater than 1 from a specified state of the flow table. If the above circuit was to be realized with clocked-D flip-flops, then $D_i = Y_i$ in the above equations.

There are many lurking faults associated with this design procedure. For example, six of the eight product terms in Y_4 could be s-a-0 and the fault conditions would be undetectable since they only affect states that the circuit normally does not enter. As in the asynchronous case where a minimum distance code was used, the logic that provides for fault tolerance could fail and not yield any observable change in circuit behavior and give the appearance of being able to tolerate a single fault. When in effect, the next fault that does occur may force the circuit to malfunction. The disadvantages associated with using a minimum distance 3 code in the asynchronous case apply to the synchronous case.

One procedure found to be effective in designing synchronous sequential circuits, when viewed from a practical standpoint, is that depicted in Fig. 12, where fault detection is achieved by using the 1-out-of-n code described earlier. The advantages of this procedure are (1) it is easy to design each fault detecting circuit, (2) it is more efficient in hardware (even though a proof of this conjecture is not given), and (3) it does not suffer from lurking fault conditions like the minimum distance 3 code approach.

REFERENCES

1. A. L. Hopkins, Jr., and T. B. Smith, III, "The Architectural Elements of a Symmetric Fault-Tolerant Multiprocessor," Digest of Fourth International Symposium on Fault Tolerant Computing, pp. 4-2, 4-6, June 1974.

2. S. H. Unger, *Asynchronous Sequential Switching Circuits*, New York, Wiley-Interscience, 1969.

3. W. W. Patterson, and G. A. Metze, "A Fault-Tolerant Asynchronous Sequential Machine," 1972 International Symposium on Fault-Tolerant Computing, pp. 176-181, June 1972.

4. D. K. Pradhan, and S. M. Reddy, "Fault-Tolerant Asynchronous Networks," *IEEE Trans. Comput.*, vol. C-22, pp. 662-668, July 1973.

5. D. H. Sawin, III, and G. K. Maki, "Asynchronous Sequential Machines Designed for Fault Detection," *IEEE Trans. Comput.*, vol. C-23, pp. 239-249, March 1974.

6. G. K. Maki, and D. H. Sawin, III, "Fault-Tolerant Asynchronous Sequential Machines," *IEEE Trans. Comput.*, vol. C-23, pp. 651-657, July 1974.

7. D. H. Sawin, III, and G. K. Maki, "Failsafe Asynchronous Circuits," *IEEE Trans. Comput.*, Vol. C-24, pp. 675-677, June 1975.

8. J. F. Meyer, "Fault-Tolerant Synchronous Sequential Machines," *IEEE Trans. Comput.*, Vol. C-20, October 1971.

9. R. L. Russo, "Synthesis of Error-Tolerant Counters Using Minimum Distance Three State Assignments," *IEEE Trans. on Electronic Computers*, vol. EC-14, June 1965.

SOME METHODS FOR ON-LINE DETECTION AND/OR DIAGNOSIS OF CONTINUOUS LINEAR SYSTEMS: A SURVEY

R. F. Garzia

Introduction

In the last fifteen years more than 150 technical papers have been written in the general area of detection and/or diagnosis of continuous systems. A big percentage of them have been developed under sponsorship of Wright-Patterson Air Force Base, Ohio. A complete bibliography is included at the end of this paper.

The main objective for the search of all the fault isolation methods has been to identify the mathematical techniques used in such a purpose. Of particular interest were those amenable for computer application. The different techniques individualized during this study are presented in references [46] and [51].

Within the scope of the present paper we are interested in the mathematical techniques that allow the verification of the status of the system under test in an on-line basis. Furthermore, we would like to accomplish such a task using only the normal inputs/outputs terminals. Or in other words, we do not want the addition of special terminals for testing purposes.

Let us clarify the above idea. Doing this to our system under test called SUT (hereafter) will take the appearance shown in Figure 1. We have a finite number of inputs and outputs, and the system itself is subject to some disturbances. The measurements that we will take are obtained from the inputs/outputs through sensors and/or filters. These units are also subject to disturbances.

One of our purposes, also, will be to minimize the effect of disturbances in the results. We will call the measured inputs and outputs the observables of the system. Therefore, our problem is stated to obtain the system status knowing the observables of it.

Since we are concerned with accomplishing the above statement in an on-line basis, the configuration needed will look like the one shown in Figure 2.

At this time it becomes important to define what we mean by "detection," "diagnosis," "system healthy," and "system sick". Let us start

R. F. Garzia is Senior Application Analyst with the Babcock & Wilcox Company in Barberton, Ohio.

by saying that in the process of testing a given system, we are going to distinguish two processes:

- Detection
- Diagnosis

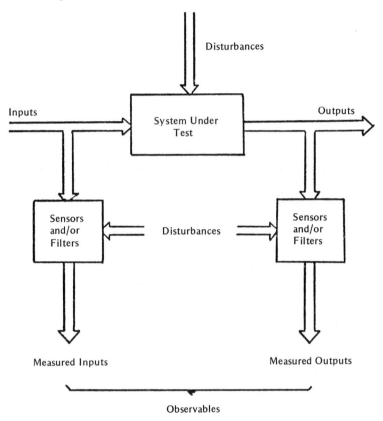

Figure 1

The <u>Detection Process</u> is the process of detecting a fault in the SUT, disregarding the cause of this fault. This could be a single or a multiple fault. We see that in this process we are concerned only with whether the system is healthy or sick, and not with the cause of its sickness.

The <u>Diagnosis Process</u> is the process of detecting what is the cause or causes of sickness. Or in other words it performs the diagnosis of the fault.

The above definitions suggest that the normal sequence of operations will be to perform first the detection process and then, if the system is sick, perform the diagnostic one. Although this sequence is the normal one, in some mathematical techniques both processes are simultaneously performed.

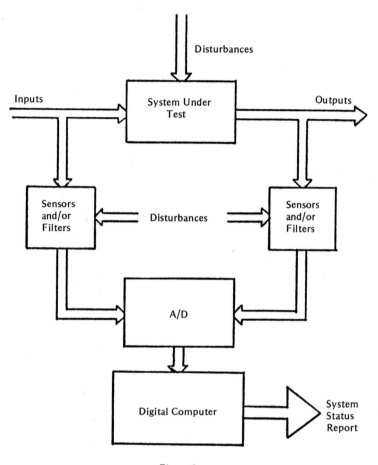

Figure 2

In a general sense, the flowchart shown in Figure 3 gives us the different alternatives involved in these processes.

Since our main concern is with linear systems, we can characterize them by means of

$$\text{Differential Equation} \longrightarrow \sum_{i=0}^{n} a_i \frac{d^i f(t)}{dt^i} = \sum_{i=0}^{m} b_i \frac{d^i u(t)}{dt^i} \tag{1}$$

$$\text{Difference Equation} \longrightarrow \sum_{i=0}^{n} c_i \, f((k+i)T) = 0 \tag{2}$$

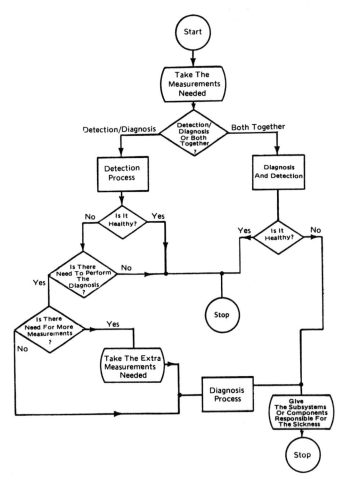

Figure 3

where $u(t)$ and $f(t)$ stands for the input and output, respectively. All the coefficients $a_i(i=0,1,2\ldots,n)$, $b_i(i=0,1,2\ldots,m)$, and $c_i(i=0,1,2\ldots,n)$ are constants.

By definition, we will assume that a system represented by Equation (1) is healthy if all the coefficients a_i and b_i satisfy

$$\left\{ \begin{array}{ll} a_{imin} \leq a_i \leq a_{imax} & i=0,1,2,\ldots,n \\ b_{imin} \leq b_i \leq b_{imax} & i=0,1,2,\ldots,m \end{array} \right. \tag{3}$$

where $a_{imin} - a_{imax}$ and $b_{imin} - b_{imax}$ represents the normal operational range.

If the coefficients a_i and b_i do not satisfy Inequalities (3) our system is sick. Similarly, this holds for the coefficients represented by Equation (2). Or, the system is healthy if

$$c_{imin} \leqslant c_i \leqslant c_{imax} \qquad\qquad i=0,1,2, \ldots n \qquad\qquad (4)$$

Otherwise it is sick.

The techniques selected within this study are:

- Detection Technique
- Impulse Response
- Linear Programming
- RLL Transform

Within each technique the mathematical background is presented as well as a simple application.

Detection Technique [46, 47, 48, 49]

We assume that the curve $h_n(t)$ shown in Figure 4 represents the theoretical impulse response of the SUT. For any values of the system's components, other than the theoretical ones, we are going to have the impulse response shown as $h(t)$. The samples obtained from both curves must

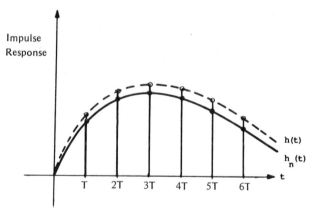

Figure 4

satisfy Equation (2). Then

$$\sum_{i=0}^{n} c_i \, h_n((n+i)T) = 0$$

$$\sum_{i=0}^{n} c_i' \, h((n+i)T) = 0 \qquad\qquad (5)$$

and we can write

$$c'_i = c_i + \epsilon_i \tag{6}$$

Therefore, the second of Equations (5) can be written as follows

$$\sum_{i=0}^{n} c_i\, h((n+i)T) + \sum_{i=0}^{n} \epsilon_i\, h((n+i)T) = 0 \tag{7}$$

Now, if in Equation (7) we can write ϵ_{imax} instead of ϵ_i, as the maximum allowed error in the ith coefficient, we obtain the inequality that follows

$$\sum_{i=0}^{n} c_i\, h((n+i)T) + \sum_{i=0}^{n} \epsilon_{imax}\, h((n+i)T) \geqslant 0 \tag{8}$$

Similarly with the minimum error allowed

$$\sum_{i=0}^{n} c_i\, h((n+i)T) - \sum_{i=0}^{n} \epsilon_{imin}\, h((n+i)T) \leqslant 0 \tag{9}$$

if $\epsilon_{imin} = \epsilon_{imax} = \epsilon_{im}$ we have

$$\left| \sum_{i=0}^{n} c_i\, h((n+i)T) \right| \leqslant \sum_{i=0}^{n} \epsilon_{im} |h((n+i)T)| \tag{10}$$

The above condition is a necessary but insufficient one for all the coefficients of the system to be within their allowed range. However if

$$\left| \sum_{i=0}^{n} c_i\, h((n+i)T) \right| > \sum_{i=0}^{n} \epsilon_{im} |h((n+i)T)| \tag{11}$$

it is a sufficient condition for one or more coefficients to be out of the allowed range. It is possible to define a scalar performance indicator as follows

$$\delta(nT) = \left| \sum_{i=0}^{n} c_i\, h((n+i)T) \right| - \sum_{i=0}^{n} \epsilon_{im} |h((n+i)T)| \tag{12}$$

Then, if $\delta(nT) \geqslant 0$ we know for sure that at least one coefficient is out of the allowed range. On the other hand, the condition $\delta(nT) < 0$ is a necessary but not sufficient one for all the coefficients lying within the allowed range. The preceding statement is true if all the c_i coefficients are not related.

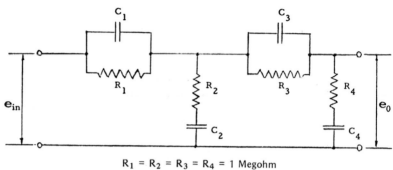

$$R_1 = R_2 = R_3 = R_4 = 1 \text{ Megohm}$$
$$C_1 = C_2 = C_3 = C_4 = 1 \, \mu \text{ Farad ay}$$

$T_1 = R_1 C_1$	$T_{12} = R_1 C_2$
$T_2 = R_2 C_2$	$T_{13} = R_1 C_3$
$T_3 = R_3 C_3$	$T_{14} = R_1 C_4$
$T_4 = R_4 C_4$	$T_{34} = R_3 C_4$

Figure 5

But in a different equation that represents a physical system some relationship exists between them. Therefore, for these cases the condition $\delta(nT) < 0$ is necessary and in general is sufficient.

As an example, let us take the circuit shown in Figure 5. Assuming that all the components are in the design values and only R_1 is allowed to vary in the range $0.4R_1$ to $2.5R_1$ we obtain the curves shown in Figure 6. The three curves given are from different values of ϵ_{im}. The upper one is for $\epsilon_{im} = 0.0$; $i = 1, 2, \ldots 8$. Similarly for the other components. For this application, the condition $\delta(nT) < 0$ is necessary and sufficient for a healthy system.

Impulse Response [36]

The basic idea of this method is the sampling of the output of the system when it is excited with a unit impulse. From these values, we find the coefficients of the transfer function. The applicability of this method in this way is constrained by the following facts:

- Practical unit impulse with small width in order to drive the system.

- Low energy level of system excitation given by the small area of unit impulse.

- Return to zero of the pulse.
- Repeatability of the pulse.

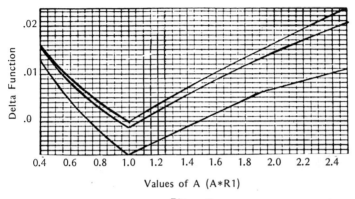

Values of A (A∗R1)

Figure 6

To avoid these problems, autocorrelation techniques can be used in an on-line basis with any kind of excitations. See reference [8].

Briefly the mathematical background is as follows: Taking the Laplace Transform of Equation (1) we have

$$H(s) = \frac{K_1}{s-s_1} + \frac{K_2}{s-s_2} + \cdots + \frac{K_n}{s-s_n} = \sum_{i=1}^{n} K_i \frac{1}{s-s_i} \qquad (13)$$

where,

$$K_i = \left| \frac{c_n z^{n-1}\langle f(kT) + \cdots + f((k+n-1)T)z^{-n+1}\rangle + \cdots + c_n f(kT)}{(z-z_1) \cdots (z-z_{i-1})(z-z_{i+1}) \cdots (z-z_n)} \right|_{z=z_i}$$

$$i=1,2,\ldots,n$$

$$(14)$$

The steps involved in the method are as follows:

1. Take the measurements for 2n equal spaced samples, where n is the degree of the impulse response of the SUT. These output samples are given by

$$f(kT); f((k+1)T); \ldots ; f((k+2n-1)T) \qquad (15)$$

2. Find the values of the a's constants from the system of equations that follows

$$\begin{cases} a_n f((k+n)T) & +a_{n-1}f((k+n-1)T) & + \cdots +a_1\, f((k+1)T) = -f(kT) \\ a_n f((k+n+1)T) & + a_{n-1}f((k+n)T) & + \cdots + a_1\, f((k+2)T) = -f((k+1)T) \\ \cdots\cdots\cdots\cdots\cdots\cdots\cdots\cdots\cdots\cdots\cdots\cdots\cdots\cdots\cdots\cdots\cdots\cdots\cdots\cdots \\ a_n\, f((k+2n-1)T) & + a_{n-1}f((k+2n-2)T) & + \cdots + a_1\, f((k+n)T) \end{cases}$$

$$= -f((k+n-1)T) \qquad (16)$$

3.   Find the values of the roots of equation

$$a_n z^n + a_{n-1}z^{n-1} + \cdots + a_1 z + 1 = 0$$

and designate them with $z_1, z_2, \ldots, z_n$. $\qquad\qquad\qquad (17)$

4.   Find the values of the poles of our system by means of the equation

$$s_i = \frac{1}{T}\ln z_i \quad (i=1,2,\, \cdots, n) \qquad\qquad (18)$$

5.   Find the values of the K's coefficients by means of Equation (14) and the transfer function by Equation (15).

As an example, let us assume that from a fourth order system, having a sampling period of 0.01 seconds, we obtain the values of the impulse response that follows

Time	Input	Output
0.10	1.00	0.0000
0.11	1.00	0.4116
0.12	1.00	0.7740
0.13	1.00	1.0917
0.14	1.00	1.3690
0.15	1.00	1.6096
0.16	1.00	1.8171
0.17	1.00	1.9946

The values of the a's coefficients are obtained from

$$\begin{bmatrix} 1.3690 & 1.0917 & 0.7740 & 0.4116 \\ 1.6096 & 1.3690 & 1.0917 & 0.7740 \\ 1.8171 & 1.6096 & 1.3690 & 1.0917 \\ 1.9946 & 1.8171 & 1.6096 & 1.3690 \end{bmatrix} \begin{bmatrix} a_1 \\ a_2 \\ a_3 \\ a_4 \end{bmatrix} = \begin{bmatrix} 0.0000 \\ -0.4116 \\ -0.7740 \\ -1.0917 \end{bmatrix}$$

Then, the a_i values are

$$a_1 = 1.2297; a_2 = -4.6719; a_3 = 6.6551; a_4 = -4.2129$$

Knowing these values we form the polynomial

$$1.2297z^4 - 4.6719z^3 + 6.6551z^2 - 4.2129z + 1 = 0$$

the roots of which are

$$z_1 = 0.9762; z_2 = 0.9544; z_3 = 0.9377; z_4 = 0.9308$$

And the poles and K coefficients of our system are

$$s_1 = -2.419; s_2 = -4.656; s_3 = -6.453; s_4 = -7.181$$

$$K_1 = -1.501; K_2 = 21.234; K_3 = 10.132; K_4 = -29.645$$

The coefficients of the transfer function are calculated from the above values.

Linear Programming [40]

The basic idea is as follows: "Given the input and output of a process as a sequence of sampled-data points, linear programming can be used to obtain the values of the transfer function coefficients that describe the process in question".

Let us assume that our system looks like the one shown in Figure 7.

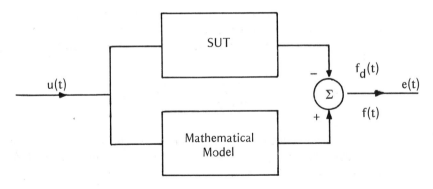

Figure 7

We assume that the transfer function of the SUT is given by the ratio of two polynomials, or

$$H(s) = \frac{N(s)}{D(s)} = \frac{c_m s^m + c_{m-1} s^{m-1} + \cdots + c_1 s + c_0}{s^n + d_{n-1} s^{n-1} + \cdots + d_1 s + d_0} \qquad (19)$$

where $n > m$.

Now let us define the error function as
$$e(t) = f(t) - f_d(t) \tag{20}$$
If we take the samples at K different values of time, we want to find the coefficients of Equation (19), such that
$$\sum_{k=1}^{K} |e(kT)| \tag{21}$$
will be a minimum. We would like to use Linear Programming for this purpose. To do this, some modifications need to be made on Equation (19). This particular modification can be accomplished by the introduction of an iteration procedure [110]. In doing so let us write the Laplace Transform of Equation (20).

$$E(s) = F(s) - F_d(s) \tag{22}$$

Taking into consideration Equation (19), we can write the above equation as follows
$$E(s) = \frac{N(s)}{D(s)} U(s) - F_d(s) \tag{23}$$

or,

$$E(s) = \frac{N(s)}{D(s)} U(s) - \frac{D(s)}{D(s)} F_d(s) \tag{24}$$

Let us assume that both polynomials' numerators in Equation (24) are estimated in the ith iteration, and the denominator is estimated in the previous iteration. Under these conditions we can write Equation (24) as follows

$$E_i(s) = \frac{N_i(s)}{D_{i-1}(s)} U(s) - \frac{D_i(s)}{D_{i-1}(s)} F_d(s) \tag{25}$$

Note that if the estimates of the numerators, $N_i(s)$ and $D_i(s)$, converge in their true values, then $D_i(s)$ asymptotically equals $D_{i-1}(s)$ and Equation (25) becomes Equation (23). Now let

$$U'(s) = \frac{U(s)}{D_{i-1}(s)} \; ; \; F'_d(s) = \frac{F_d(s)}{D_{i-1}(s)} \tag{26}$$

Therefore, Equation (25) becomes

$$E_i(s) = N_i(s)U'(s) - D_i(s)F'_d(s) \tag{27}$$

and this suggests the system shown in Figure 8.

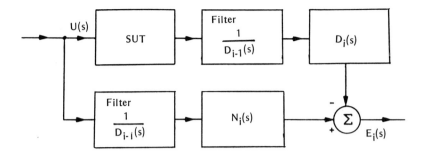

Figure 8

Expanding the polynomials indicated in Equation (27) we have

$$E_i(s) = [c_m s^m + c_{m-1} s^{m-1} + \cdots + c_1 s + c_0] U'(s)$$
$$-[s^n + d_{n-1} s^{n-1} + \cdots + d_1 s + d_0] F_d'(s) \qquad (28)$$

Therefore, in the time domain Equation (28) becomes

$$e_i(t) = c_m q_{cm}(t) + c_{m-1} q_{c(m-1)}(t) + \cdots + c_1 q_{c1}(t) + c_0 q_{c0}(t)$$
$$-[q_{dn}(t) + d_{n-1} q_{d(n-1)}(t) + \cdots + d_1 q_{d1}(t) + d_0 q_{d0}(t)]$$

where $\qquad (29)$

$$q_{cj}(t) = \mathcal{L}^{-1} \left[\frac{s^j U(s)}{D_{i-1}(s)} \right] \qquad j = 0, 2, \ldots, m$$
$$q_{dj}(t) = \mathcal{L}^{-1} \left[\frac{s^j F_d(s)}{D_{i-1}(s)} \right] \qquad j = 0, 1, \ldots, (n-1) \qquad (30)$$

Let us assume for the moment that all the above functions are known. Under these conditions, if we replace the continuous function of t by kT, where k=1,2, . . .K, we have the following system of equations

$$e_i(T) = c_m q_{cm}(T) \cdots + c_0 q_{co}(T) - q_{dn}(T) \cdots - d_0 q_{do}(T)$$
$$e_i(2T) = c_m q_{cm}(2T) \cdots + c_0 q_{co}(2T) - q_{dn}(2T) \cdots - d_0 q_{do}(2T)$$
$$\cdots\cdots\cdots\cdots\cdots\cdots\cdots\cdots\cdots\cdots\cdots\cdots\cdots\cdots$$

$$e_i(KT) = c_m q_{cm}(KT) \cdots + c_0 q_{co}(KT) - q_{dn}(KT) \cdots - d_0 q_{do}(KT)$$

$$\qquad (31)$$

Now since we have the following ranges of values for the unknown variables involved in the above equations

$$-\infty < e_i(kT) < \infty$$
$$-\infty < c_j < \infty$$
$$0 \leqslant d_j < \infty$$

we can write

$$\begin{cases} e_i (kT) \rightarrow e_i^+(kT) - e_i^- (kT) \\ c_j \rightarrow c_j^+ - c_j^- \end{cases} \tag{32}$$

where these new variables take only positive values. Under these conditions Equations (31) become

$$e_i^+(T)-e_i^-(T) = c_m^+ q_{cm}(T)-c_m^- q_{cm}(T)+\cdots + c_o^+ q_{co}(T)-c_o^- q_{co}(T) + \cdots$$
$$e_i^+(2T)-e_i^-(2T) = c_m^+ q_{cm}(2T)-c_m^- q_{cm}(2T)+\cdots+ c_o^+ q_{co}(2T)-c_o^- q_{co}(2T) + \cdots$$
$$\cdots\cdots\cdots\cdots\cdots\cdots\cdots\cdots\cdots\cdots\cdots\cdots\cdots\cdots\cdots\cdots$$
$$e_i^+(kT)-e_i^-(KT)= c_m^+ q_{cm}(KT)-c_m^- q_{cm}(KT) + \cdots +c_o^+ q_{co}(KT)-c_o^- q_{co}(KT) + \cdots$$

$$-[q_{dn}(T)+d_{n-1}q_{d(n-1)}(T) +\cdots +d_o q_{do}(T)]$$
$$-[q_{dn}(2T)+d_{n-1}q_{d(n-1)}(2T) +\cdots +d_o q_{do}(2T)]$$
$$\cdots\cdots\cdots\cdots\cdots\cdots\cdots\cdots\cdots\cdots\cdots\cdots$$
$$-[q_{dn}(KT)+d_{n-1}q_{d(n-1)}(KT) +\cdots +d_o q_{do}(KT)] \tag{33}$$

Therefore, within each iteration, our linear programming can be stated as follows

$$\text{Minimize} \sum_{k=1}^{K} [e_i^+ (kT) + e_i^- (kT)] \tag{34}$$

subject to

$$e_i^+(T)-e_i^-(T) +\cdots -c_m^+ q_{cm}(T)+c_m^- q_{cm}(T) + \cdots$$
$$+ e_i^+(2T) -e_i^-(2T) +\cdots -c_m^+ q_{cm}(2T)+c_{mm}^- q_{cm}(2T) + \cdots$$
$$\cdots\cdots\cdots\cdots\cdots\cdots\cdots\cdots\cdots\cdots\cdots\cdots\cdots\cdots\cdots$$
$$+ e_i^+(KT)-e_i^- (KT) -c_m^+ q_{cm}(KT)+c_m^- q_{cm}(KT) + \cdots$$

$$-c_o^+ q_{co}(T)+c_o^- q_{co}(T)+d_{n-1}q_{d(n-1)}(T) +\cdots +d_o q_{do}(T) = q_{dn}(T)$$
$$-c_o^+ q_{co}(2T)+c_o^- q_{co}(2T)+d_{n-1}q_{d(n-1)}(2T) +\cdots +d_o q_{do}(2T) = q_{dn}(2T)$$
$$\cdots\cdots\cdots\cdots\cdots\cdots\cdots\cdots\cdots\cdots\cdots\cdots\cdots\cdots\cdots$$
$$-c_o^+ q_{co}(KT)+c_o^- q_{co}(KT)+d_{n-1}q_{d(n-1)}(KT) +\cdots +d_o q_{do}(KT) = q_{dn}(KT)$$
$$\tag{35}$$

No matter what kind of functions $f(t)$ and $f_d(t)$ are involved with, they may be approximated as the sum of a series of rump functions.

A complete application of this method is presented in reference [40].

RLL Transform [53, 54, 118, 119].

The Running Leftside Laplace Transform is given by

$$F(s,t) = \int_{t_s}^{t} f(v) \, e^{s(t-v)} dv \qquad (36)$$

And in general for the RLL transform of the nth derivative, we can write

$$\int_{t_s}^{t} \frac{d^n f(v)}{dt^n} \, e^{s(t-v)} dv = s^n F(s,t) + \sum_{i=0}^{n-1} s^{n-i-1} \frac{d^i f(t)}{dt^i} \qquad (37)$$

The introduction of the RLL transform allows the generalization of the transfer function. To define such transfer function, let us take under consideration the system shown in Figure 9.

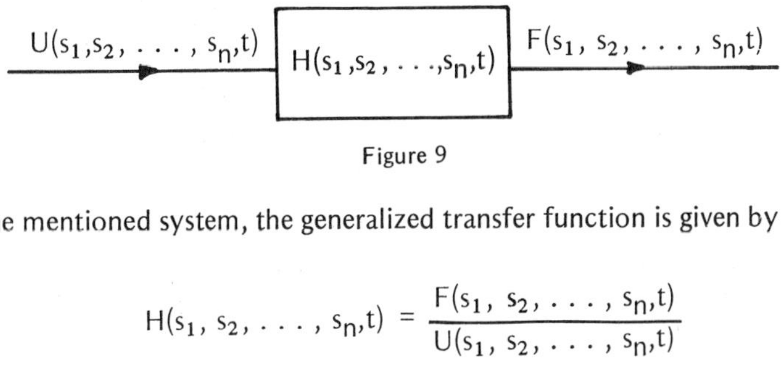

Figure 9

For the mentioned system, the generalized transfer function is given by

$$H(s_1, s_2, \ldots, s_n, t) = \frac{F(s_1, s_2, \ldots, s_n, t)}{U(s_1, s_2, \ldots, s_n, t)}$$

$$= \frac{[b_n s_1^m + b_{m-1} s_1^{m-1} + \cdots + b_1 s_1 + b_0] + \sum_{i=1}^{m} \frac{U(s_i+1, \ldots, s_n, t)}{U(s_1, s_2, \ldots, s_n, t)}[\text{polynomial coeff.}]}{[s_1^n + a_{n-1} s_1^{n-1} + \cdots + a_1 s_1 + a_0] + \sum_{i=1}^{m} \frac{F(s_i+1, \ldots, s_n, t)}{F(s_1, s_2, \ldots, s_n, t)}[\text{polynomial coeff.}]}$$

$$(38)$$

where if

$$\begin{cases} i+1 = n & Y(s_i+1, \ldots, s_n,t) = Y(s,t) \\ i = n & Y(s_i+1, \ldots, s_n,t) = Y(t) \end{cases} \tag{39}$$

    A complete deduction of Equation (38) as well as the formation rule of the polynomial coefficients is given in References [54] and [55]. To measure the variables required in Equation (38), we can use the arrangement shown in Figure 10. We can simplify the measurements in this way because our system is linear and therefore, we can reverse the transformation order, or

$$F(s_1, s_2, \ldots, s_n,t) = F(s_n, s_{n-1}, \ldots, s_1,t) \tag{40}$$

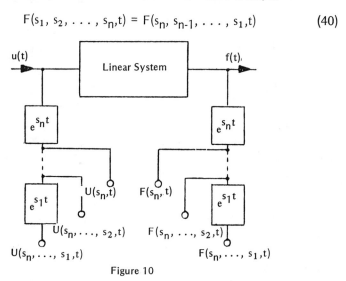

Figure 10

    Since the RLL transforms are functions of the complex variables $s_1, s_2, \ldots, s_n$ as well as of t, to find the values of the coefficients of the generalized transfer function given by Equation (38) two approaches can be used

    1. Variable Time
    2. Fixed Time

    Both approaches are completely discussed in reference [55]. At this time we would concentrate only on the one using the RLL transform with linear programming. In this condition the generalization of Equation (25) becomes

$$E_i(s_1,s_2, \ldots, s_n,t) = \frac{N_i(s_1,s_2, \ldots, s_n,t)}{D_{i-1}(s_1,s_2, \ldots, s_n,t)} U(s_1,s_2, \ldots, s_n,t)$$

$$-\frac{D_i(s_1,s_2,\ldots,s_n,t)}{D_{i-1}(s_1,s_2,\ldots,s_n,t)}F(s_1,s_2,\ldots,s_n,t) \tag{41}$$

See reference [53].

In general for the generalized transfer function given by Equation (38) we can write

$$H(s_1,s_2,\ldots,s_n,t) = \frac{\sum\limits_{j=0}^{m} b_j\,\beta_j(s_1,s_2,\ldots,s_n,t)}{\sum\limits_{j=0}^{n} a_j\,a_j(s_1,s_2,\ldots,s_n,t)} \tag{42}$$

where

$$\beta_0(s_1,s_2,\ldots,s_n,t) = a_0(s_1,s_2,\ldots,s_n,t) = a_n = 1$$

Now Equation (41) becomes

$$E_i(s_1,s_2,\ldots,s_n,t) = \sum_{j=0}^{m} b_j \left[\frac{\beta_j(s_1,s_2,\ldots,s_n,t)}{D_{i-1}(s_1,s_2,\ldots,s_n,t)}\,F(s_1,s_2,\ldots,s_n,t)\right] \tag{43}$$

$$-\sum_{j=0}^{n} a_j \left[\frac{a_j(s_1,s_2,\ldots,s_n,t)}{D_{i-1}(s_1,s_2,\ldots,s_n,t)}\,Y(s_1,s_2,\ldots,s_n,t)\right]$$

let

$$\left\{\begin{array}{l} \theta_j^{i-1}(s_1,s_2,\ldots,s_n,t) = \dfrac{\beta_j(s_1,s_2,\ldots,s_n,t)}{D_{i-1}(s_1,s_2,\ldots,s_n,t)}\,F(s_1,s_2,\ldots,s_n,t) \\[4mm] \lambda_j^{i-1}(s_1,s_2,\ldots,s_n,t) = \dfrac{a_j(s_1,s_2,\ldots,s_n,t)}{D_{i-1}(s_1,s_2,\ldots,s_n,t)}\,Y(s_1,s_2,\ldots,s_n,t) \end{array}\right. \tag{44}$$

Therefore Equation (43) becomes

$$E_i(s_1,s_2,\ldots s_n,t) = \sum_{j=0}^{m} b_j \theta_j^{i-1}(s_1,s_2,\ldots s_n,t)$$
$$-\sum_{j=0}^{n} a_j \lambda_j^{i-1}(s_1,s_2,\ldots s_n,t) \qquad (45)$$

To find the values of the coefficients of Equation (45), two approaches can be used: Fixed Time and Variable Time.

Within the *fixed time* approach we can form K linear equations assigning a different set of values for the complex variables $s_1, s_2, \ldots s_n$ and taking all the measurements at the same time. Equation (45) becomes

$$E_i(s_1^k,s_2^k,\ldots,s_n^k,t) = \sum_{j=0}^{m} b_j \theta_j^{i-1}(s_1^k,s_2^k,\ldots,s_n^k,t) - \sum_{j=0}^{n} a_j \lambda_j^{i-1}(s_1^k,s_2^k,\ldots,s_n^k,t)$$

for $k = 0, 1, \ldots, K-1$

$$(46)$$

For the *variable time* approach, it is convenient to replace the continuous function of t by kT, where T is the sampling period and $k=0,1,2,\ldots K-1$. Under this condition Equation (45) becomes

$$E_i(s_1,s_2,\ldots,s_n,kT) = \sum_{j=0}^{m} b_j \theta_j^{i-1}(s_1,s_2,\ldots,s_n,kT) - \sum_{j=0}^{n} a_j \lambda_j^{i-1}(s_1,s_2,\ldots,s_n,kT)$$

$$(47)$$

Although both approaches can be used as well as a combination of them, we pursue the second one because:

- It requires fewer networks for implementation
- It is less sensitive to noise
- For a given interval of time the maximum error could be bounded
- It requires fewer calculations

Assuming that all the complex variables are fixed in real values and $a_n = 1$, Equation (47) becomes

$$-E_i(kT) + \sum_{j=0}^{m} b_j \theta_j^{i-1}(kT) - \sum_{j=0}^{n-1} a_j \lambda_j^{i-1}(kT) = \lambda_n^{i-1}(kT) \qquad (48)$$

Assuming that we have stable systems, the signs associated with each variable of Equation (48) are

$$\begin{array}{lll} E_i(kT) & k = 0, 1, 2, \ldots, k-1 & + \text{ or } - \\ b_j & j = 0, 1, 2, \ldots, m & + \text{ or } - \\ a_j & j = 0, 1, 2, \ldots, n-1 & + \end{array}$$

Therefore, we need to write

$$\begin{cases} E_i(kT) &= E_i^{\;+}(kT) - E_i^{\;-}(kT) \\ b_j &= b_j^{\;+} - b_j^{\;-} \end{cases} \tag{49}$$

Under these conditions, the number of variables of our linear programming is given by

$$\text{Number of Variables} = 2(K+m) + n + 2 \tag{50}$$

We can write that the objective function of our problem is

$$\text{Minimize} \sum_{k=0}^{K-1} [E_i^{\;+}(kT) + E_i^{\;-}(kT)] \tag{51}$$

subject to the equality constraints

$$-E_i^{\;+}(0) \quad +E_i^{\;-}(0) \quad + \sum_{j=0}^{m} (b_j^{+} - b_j^{-}) \, \theta_j^{i-1}(0) - \sum_{j=0}^{n-1} a_j \lambda_j^{i-1}(0) \quad = \lambda_n^{i-1}(0)$$

$$\cdots\cdots\cdots\cdots\cdots\cdots\cdots\cdots\cdots\cdots\cdots\cdots\cdots\cdots\cdots\cdots\cdots\cdots \tag{52}$$

$$-E_i^{\;+}((K-1)T) + E_i^{\;-}((K-1)T) + \sum_{j=0}^{m} (b_j^{+} - b_j^{-}) \theta_j^{i-1}((K-1)T) - \sum_{j=0}^{n-1} a_j \lambda_j^{i-1}((K-1)T) = \lambda_n^{i-1}((K-1)T)$$

Let us apply the approach already presented to a specific problem. In doing so we will assume that the system is described by the ordinary differential equation that follows

$$\frac{d^2 y(t)}{dt^2} + 5\frac{dy(t)}{dt} + 6\,y(t) = 2\frac{df(t)}{dt} + 3f(t)$$

with: $s_1 = -2.0$ and $s_2 = -3.0$

Without noise superimposed on measurements, we have

$K = 8$
Starting Values $a_0 = 2.0; \; a_1 = 1.0$
Two Iterations $a_0 = 5.9819; \; a_1 = 4.9987$
                                         $b_0 - 2.9910; \; b_1 = 2.0010$

With 0.05% of noise superimposed on measurements

$$K = 20$$

Starting Values $a_0 = 2.0; a_1 = 1.0$

Two Iterations $a_0 = 5.5583; a_1 = 4.9031$

                                $b_0 = 2.7752; b_1 = 1.9865$

Methods Comparison

The properties that follow will be weighed in this evaluation:

1. Post calculations on measurements— This implies only the calculations needed to be performed on the measurements taken in order to be able to apply the method in question.

2. Influence of noise on measurements— This covers the influence on the results of the noise added on measurements.

3. Low order models—This stands for the capability of the method that allows us to get low order mathematical models.

4. Possibility of one time measurements— This stands for the capability of taking all the information required for system identification at only one time.

5. Error constrained—This counts for the capability of constraining the errors at specific values of time. The error is defined as the difference between the response of the mathematical model and the response of the actual system.

The results of these evaluations are indicated in Table I (shown on the next page).

Conclusions

As a result of this study the following conclusions can be stated:

- For lumped systems the RLL transform method presents strong advantages over the others.

- Combination of methods improves the results of the SUT and minimizes disturbances, i.e., RLL transform combined Linear Programming.

- Solution of the generalized transfer function is stated in a compact matrix form that suggests the possibility of preparing standard programs.

- More work must be done in this area to have a complete evaluation of RLL transform method in different types of applications.

TABLE I — METHODS COMPARISON

Method	Post Calculations on measurements	Influence of noise on measurements	Low order models	Possiblity of one time measurements	Error Constrained	Most Advantageous Method
Detection Condition	No	Heavy	No	No	No	
Impulse Response	No	Heavy	No	No	No	
Linear Programming	Yes	Light	Yes	No	Yes	
RLL Transform	No	Heavy	No	Yes	No	
RLL Transform and LP	No	Light	Yes	Yes	Yes	⇐

References and Bibliography

1 Aamand, C., "Process Identification for On-Line Optimization", EUROCON71, Palais de Beaulieu Lausanne-Switzerland, October 1971.

2 Allen, R. J., "Failure Prediction Employing Continuous Monitoring Techniques", IEEE Transactions on Aerospace Support Conference Proc., 1963.

3 Balakrishnan, A. V., and Peterka, V., "Identification in Automatic Control Systems", Automatica, Vol. 5, pp. 817-829, 1969.

4 Barry, R. F., "Some Analytic Fault Isolation Techniques", Lecture Notes of Computer Aided Testing and Fault Identification of State Systems, The University of Wisconsin, May 23, 1968.

5 Barry, R. F., "Fault Isolation by Parameter Identification", Listed in Automatic Support Systems Symposium, RCA, Aerospace Systems Division, October 1968.

6 Barry, R. F., Fisher, R. S., and Mattison, R. R., "Programmed Algorithm for Test Point-Selection and Fault Isolation", Vol. 1, RCA, August 1966-AD-487520.

7 Barry, R. F., Fisher, R. S., and Mattison, R. R., "Programmed Algorithm for Test Point Selection and Fault Isolation", Vol. 2, RCA, August 1966-AD-487521.

8 Baumeister, J., "In Flight Analysis of Vehicle Dynamics Along the Trajectory by Correlation Techniques", Astrionics Laboratory Marshall Space Flight Center, Hunstville, Alabama, January 1969.

9   Bayly, E. J., and Leradi, V. S., "A Method of Dynamic System Testing", Proc. Eighth Conference on Military Electronics, September 1964.

10   Beckman, C., Adler, J., Bedrosian, S. D., Berkowitz, R. S., and Chen, T. C., "Study of Piece Part Isolation of Computer Logic", Pennsylvania University, Vol. 4, June 1964-AD-136615.

11   Beckman, C., Adler, J., Bedrosian, S.D., Berkowitz, R. S., and Chen, T. C., "Study of Piece Part Isolation of Computer Logic", Pennsylvania University, Vol. 5, June 1964-AD-115337.

12   Berkowitz, R. S., and Wexelbrat, R. L., "Statistical Considerations in Element Value Solutions", IRE Transactions on Military Elect., 1962.

13   Berkowitz, R. S., "Conditions for Network Element Value Solvability", IRE Transactions on Circuit Theory, March 1962.

14   Berkowitz, R. S., and Krishnaswamy, "Computer Techniques for Solving Electric Circuits for Fault Isolation", IEEE Transactions ASCP.

15   Bollinger, J. G., and Bonesho, J. A., "Pulse Testing in Machine Tool Dynamic Analysis", International Journal Mach. Tool Des. Res., Vol. 5, 1965.

16   Boor, P. M., and Grimmer, R. L., "A Computerized Data Acquisition System for Real Time Fault Isolation", Lockheed Aircraft Service Co., November 1968.

17   Boxenhorn, B., "Using Kalman Filtering to Estimate Control Parameters", Control Engineering, July 1969.

18   Brown, J. M., and Lamb, J. D., "Fundamental Properties of the Impulse Response of Low-Order Linear Systems", Int. Journal of Control, Vol. 9, No. 2, 1969.

19   Brown, F. D., McAllister, N. F., and Perry, R. P., "An Application of Inverse Probability to Fault Isolation", IRE Trans. Military Elect.

20   Buschsbaum, L., Dunning, M., Hannom, T. J. B., and Math, L., "Investigation of Fault Diagnosis by Computational Methods", Remington Rand, May 1964-AD-601204.

21   Cabra, D., "On-Board Checkout System Concept and Philosophy", NASA Manned Space Center, Automatic Support System for Advanced Maintainability, Nov., 1968.

22   Chesler, L. G., and Turn, R., "Some Aspects of Man-Computer Communications in Active Monitoring of Automated Checkout", The Rand Corp., March 1967-AD-648553.

23   Chesler, L., and Turn, R., "The Monitoring Task in Automated Checkout of Space Vehicles", The Rand Corporation, Memorandum RM-4678-NASA-Sep. 1965.

24   Chorafas, D. N., "Systems and Simulation", Washington State University, Mathematics in Science and Engineering, A. P., Vol. 14, 1965.

25   Constanza, J. L., and Osborne, R. L., "Fault Detection and Diagnosis, An Energy Point of View", Proc. of Aut. Support Syst. Symp. for Adv. Maintainability, June 1965.

26   Cox, H., "On the Estimation of State Variables and Parameters for Noisy Dynamic Systems", IEEE Transactions on A. C., January 1964.

27   Cuenod, M., and Sage, A. P., "Comparison of Some Methods Used for Process Identification", Automatica, Vol. 4, No. 4, pp. 235-269, 1968.

28   Curry, E. R., and Schweppe, F. C., "Introduction to the Application of Modern Estimation Techniques". Special Summer Program 16.375, 1970 Course Note, Massachusetts Institute of Technology.

29 Curtis, G. C., and Mash, J. T., "An Airborne Manual/Automatic Malfunction Detection System", Lockheed Georgia Company, Marietta, Georgia.

30 Davies, W. D. T., "System Identification for Self-Adaptive Control", Wiley Interscience, 1970.

31 Delleconne, P. E., and Capehart, B. L., "Some Modeling and Simulation Considerations in Environmental Systems Analysis", 41st National ORSA Meeting, April 1972.

32 Deitsh, R., "Systems Analysis Techniques", Prentice-Hall, 1969.

33 Dorrough, D. C., "A Methodical Approach to Analyzing and Synthesizing a Self Repairing Computer", IEEE Transactions on Computers, Volume C-18, January 1969.

34 Drezner, S. M., and Gatt, O. T., "Computer-Assisted Countdown", The Rand Corporation for NASA, Memorandum RM-4565-NASA, May 1965.

35 Dynamic Testing, "A Technique for Automatic Checkout of Closed-Loop Systems" Volume I and II, Apollo Systems Department, General Electric, Huntsville, Alabama.

36 Elsden, C. S., and Ley, A. J., "A Digital Transfer Function Analyzer Based on Pulse Rate Techniques", Automatica, Vol. 5, No. 1, pp. 51-60, 1969.

37 Even, S., and Lepel, A., "On a Problem of Diagnosis", Transactions on Circuits Theory, September 1967.

38 Eykhoff, P., "Process Parameter and State Estimation", Automatica, Vol. 4, No. 4, pp. 205-233, 1968.

39 Fault Isolation by Parameter Identification, Final Report, RCA Corporation, Defense Electronics Products, Aerospace Systems Division, Burlington, Massachusetts, CR-70-588-68, 1970.

40 Fegley, K. A., Burns, J. F., and Hollis, R. M., "Process Identification via Mathematical Programming", ORSA 41st. National Meeting, April 1972, New Orleans.

41 Finkel, S. I., Nilson, R. N., and Clair, E. S. T., "A Mathematical Automatic Fault Isolation in a Complex System", Vitro Laboratories, May 1963, AD-413305.

42 Firstman, Sidney I., and Gluss, B., "Optimum Search Routines for Automatic Fault Location", 15th Meeting ORSA, Washington D. C., May 1959.

43 Firstman, Sidney I., and Gluss, B., "Current Concepts and Issues of In-Space Support", Rand Corporation, Santa Monica, Calif., August 1964, AD-606580.

44 Fisch, S. M., and Brigida, G. R., "Computer Algorithm for Fault Isolation and Test Point Selection", RCA, Burlington, Mass., December 1964, AD-613960.

45 Furst, U. R., "Automatic Built-In Test of Advanced Avionics Systems, Part II, Design Techniques for Automatic Built In Test", TAES, July 1966.

46 Garzia, R. F., "Fault Isolation Computer Methods", NASA Contract Report, CR-1758, Computer Sciences Corporation, February 1971.

47 Garzia, R. F., "A Fast Technique for Dynamic Fault Detection", Computer Sciences Corporation, Automated Support Systems for Advanced Maintainability, October, 1970.

48 Garzia, R. F., "On the Diagnosis of Controllable and Observable Systems", 1971 International IEEE Conference on Systems, Networks and Computers, January 19-21, 1971, Oaxtepec, Morelos, Mexico.

49 Garzia, R. F., "Comparison Between Fast Techniques for Dynamic Fault Detection in Linear Systems", The Third Annual Southeastern Symposium on

System Theory, The Georgia Institute of Technology, April 5-6, 1971, Atlanta, Georgia.

50 Garzia, R. F., "COMAD-Application to Future Manned Space Vehicles", 39th National Meeting of ORSA, May 5-7, 1971, Dallas, Texas.

51 Garzia, R. F., "Dynamic Methods for Automatic Fault Detection in Continuous Linear Systems", EUROCON 71, Le Palais de Beaulieu, Lausanne, Switzerland, October 18-22, 1971.

52 Garzia, R. F., "Fault Isolation in Complex Systems via Bode Diagram Technique" 1972 IEEE International Automatic Support System Symposium for Advance Maintainability, November 13-15, 1972, Philadelphia, Pennsylvania.

53 Garzia, R. F., "System Identification Using a Modified Laplace Transform", VIIth International Congress on Cybernetics, Sep. 10-15, 1973, Namur, Belgium.

54 Garzia, R. F., "Mathematical Techniques For On-Line Determination of Transfer Function Coefficients", Seventh Asilomar Conference on Circuits, Systems and Computers, Pacific Grove, California, November 27-29, 1973.

55 Garzia, R. F., "Sensitivity Analysis of Fault Isolation Computer Methods", Part I Technical Report, Computer Sciences Corporation, NASA Contract NAS8-18405-May 1971.

56 Gluss, B., "An Optimum Policy for Detecting a Fault in a Complex System", 15th Meeting ORSA, Washington D. C., May 1959.

57 Goodman, T. P., and Reswick, J. B., "Determination of System Characteristics from Normal Operation Records", Trans. ASME 78, February 1956.

58 Goodman, D. M., "Nondestructive Testing of Electronic Circuits with Fiber Optic Scientillators Vidicon Data Sampling and Pat. Reg.", 1966.

59 Goodman, D. M., "A Review of the State of the Art in Automatic Electronic Test Equipment", Reprint Paper No. 389, International Convention, 1966.

60 Graupe, "Identification of Systems", Van Nostrand Reinhold Company, 1973.

61 Harland, G. E., Panhko, M. A., Gill, K. F., and Schwarzenback, I., "Pseudo Random Signal Testing Applied to a Diesel Engine", Contr., Vol. 13, No.128, pp. 137-140, 1969.

62 Harrison, R. W., "Identification of Linear Systems Using Mathematical Programming", Wescon Convention Record, August 1968.

63 Hay, J. K., and Blew, J. M., "Dynamic Testing and Computer Analysis of Automotive Frames", Eng. Congress, Detroit, January 1972.

64 Hoberock, L. L., and Stewart, G. W., "Input Requirements and Parametric Errors for System Identification under Periodic Excitation", Transactions of the ASME, December 1972.

65 Horrubuckle, G. D., and Spann, R. M., "Diagnosis of Single Gate Failure Combinational Circuits", IEEE Transactions on Computers, Vol. C-18, March 1968.

66 Jaegly, R. L., "Test Procedure Validation by Computer Simulation", AIAA Second Flight Test Simulation and Support Conference, Los Angeles, California, 1968.

67 Johnson, J. L., McKenzie, H. R., and Moore, D. H., "On-Board In-Flight Checkout Evaluation", IBM Federal Systems Divison, June 1968, AD-883685.

68 Johnson, R. A., Kletsky, E., and Brule, J., "Diagnosis of Equipment Failures", Syracuse University Research Institute, April 1959.

69 Johnson, J. L., "On-Board In-Flight Checkout Part I", IBM Federal Systems

Division, December 1966, AD-807534.

70 Kailath, T., "An Innovations Approach to Least-Equates Estimation, Part I: Linear Filtering in Addite White Noise", IEEE Transaction on A. C., Dec. 1968.

71 Killin, T., and Tulloss, R. E., "Automatic Test Systems", IEEE Spectrum, 1974.

72 Kirkman, R., "The Relative Effectiveness of Internally Programmed and Sequential Programmed Machine for Automatic Checkout", IEEE.

73 Kraabel, P. B., "Power Spectra Analysis as a Means of On-Line Checkout".

74 Kranton, J., and Libenson, A., "A Pattern Recognition Approach to Fault Isolation", IEEE Transactions on Aerospace Support Conference.

75 Levadi, V., and Turner, L. D., "Fault Diagnosis by White Noise Techniques", Honeywell, Inc., Minneapolis, Minnesota, May 1966.

76 Levadi, V., "Pattern Recognition Applied to Fault Detection", IEEE Conference on Military Electronics, 1965.

77 Levestein, H., "Use Difference Equations to Calculate Frequency Response from Transient Response", Control Engineering, April 1957.

78 Lohse, R., and Lauler, L., "Vade-A System for Real-Time Space Vehicle Checkout and Launch Monitoring", IEEE Transactions ASCP, 1969.

79 Lux, P. A., "An Adaptive Machine for Fault Detection of Process Control ", IEEE Transactions on Industrial Electronics and Control Instruments, December 1967.

80 Mace, A. E., Pesat, R. N., Minkoff, R. T., Wertz, J. B., and Lipis, A. H., "Formulation of System Status Control Techniques", September 1963.

81 Mast, L. T., "Prelaunch Checkout in the 1970's", The Rand Corporation, Santa Monica, California, April 1965, AD-614410.

82 Mast, L. T., "Growth of Automation in Prelaunch Checkout for Space Vehicles" The Rand Corporation, Santa Monica, California, May 1967, AD-652458.

83 Math, L., Buchsbaum, L., and Hannom, T. J. B., "Investigation of Fault Diagnosis by Computer Methods for Microcircuits", Univac, Blue Bell, Pennsylvania, November 1965, AD-623951.

84 McSweeney, K. F., "Malfunction Detection System for Advanced Spacecraft", IEEE Transaction on Aerospace and Electronics Systems, January 1966.

85 Mehra, R. K., "On-Line Identification of Linear Dynamic Systems with Applications to Kalman Filtering", Joint Automatic Control Conf., Atlanta, Georgia, June 22-26, 1970.

86 Mehra, R. K., "Identification of Stochastic Linear Dynamic Systems", IEEE Symposium on Adaptive Processes, November 1969.

87 Mehra, R. K., "On Identification of Variances and Adaptive Kalman Filtering", IEEE Transactions A. C., April 1970.

88 Moon, W. E., "Periodic Checkout and Associated Errors", IEEE Transactions on Aerospace, Vol. 2, RCA, Burlington, Massachusetts, April 1964.

89 Morse, I. E., Shapton, W. R., Brown, D. L., and Kuljanic, E., "Applications of Pulse Testing for Determining Dynamic Characteristic of Machine Tools", 13th. Inter. Machine Tool Design and Research Conference, University of Birmingham, England, September 1972.

90 Murril, P. W., Pike, R. W., and Smith, C. I., "Development of Dynamic Mathematical Models", Chem. Eng., Vol. 75, No. 19, pp. 117-120, Sep. 9, 1968.

91 Neuman, C. P., and Casasayas, G. G., "Parameter Identification Using the Calerkin

Procedure in Nonlinear Boundary-Value Problems", Transactions of the ASME, 1972.

92 NewComb, R. W., "Linear Multiport Synthesis", McGraw Hill Book Company, New York, 1966.

93 Osborne, R. L., "The Detection and Diagnosis of Malfunction in Energy Manipulating Systems", University of California, Berkeley, December 1967.

94 Phillipson, "Identification of Distributed Systems", American Elsevier Publishing Co., 1973.

95 Programming, Integration and Checkout of the Advanced Radar Traffic Control System", Univac Division of Sperry Rand Corporation, April 1965, AD-622865.

96 Puri, N. N., and Weygandt, C. N., "Transfer Function Tracking of Linear Time Varying by Means of Auxiliary Simple Lag Networks", Joint ACC, 1963.

97 Ransom, M. N., and Saeks, R., "Fault Isolation with Insufficient Measurements", Electrical Engineering Memorandum EE7207, University of Notre Dame, Notre Dame, Indiana, July 1972.

98 Reswick, J. B., "Determine System Dynamics Without Upset", Control Engineering, June 2, 1955.

99 Ricker, D. W., and Saridis, G. N., "Analog Methods for On-Line System Identification Using Noisy Measurements", Simulation, Vol. 11, No. 5, 1968.

100 Ridings, R. U., and Higgins, T. J., "Transient Response Analysis of a Class of Continuous Nonlinear Time Varying Automatic Control Systems for Functional Techniques and Multidimensional Laplace Transforms, ISA Trans., Vol. 7, pp. 166-172, 1968.

101 Saeks, R., "Fault Isolation, Component Decoupling and the Connection Groupoid", Computation Laboratory, MSFC, August 21, 1970.

102 Saeks, R., Singh, S. P., and Liu, R. W., "Fault Isolation via Component Simulation", IEEE Transaction on Circuits Theory, November 1972.

103 Sage, Melsa, "System Identification", Mathematics in Science and Technology, Edited by Richard Bellman, Vol. 80, 1971.

104 Sen, A., Sihha, N. K., and Wright, J. D., "On-Line Identification of a Dual-Input-Heat-Exchanger System", Internal Reports in Simulation, Optimization and Control, No. SOC-34, Faculty of Engineering, McMaster University, April 1974.

105 Shapiro, G., Rogers, J., and Lang. O. B., "Project FIST; Fault Isolation by Semi-automatic Techniques", IEEE Spectrum, August 1964.

106 Sinha, N. K., "On-Line Identification of Continuous Systems from Sampled Data", Internal Reports in Simulations, Optimization and Control, No. SOC-6, Faculty of Engineering, McMaster University, June 1973.

107 Sinha, N. K., and Sen, A., "On-Line System Identifications: A Critical Survey", Internal Reports in Simulation, Optimization, and Control, No. SOC-14, Faculty of Engineering, McMaster University, September 1973.

108 Stahl, W. J., "Dynamic Fault Diagnosis Techniques", Scully International, Inc., Downers Grove, Illinois, September 1968.

109 Stahl, W. J., and Maenfaa, J. H., "Development of Advanced Dynamic Fault Diagnosis Techniques", Scully International, Inc., May 1967, AD-814457.

110 Steiglitz, K., and McBride, L. E., "A Technique for the Identification of Linear Systems", IEEE Transactions on Automatic Control, October 1965.

111 Swain, A. D., and Whol, J. G., "Factors Affecting Degree of Automation in Test

and Checkout Equipment", The Rand Corporation, March 1961, AD-257887.

112 Taylor, J. W., "Automatic Checkout Systems for Combat Vehicles", Frankford Arsenal, Philadelphia, Pennsylvania, April 1964, AD-823067.

113 Teasdale, A. R., and Reynolds, J. B., "Two Ways to Get Frequency Response from Transient Data", Control Engineering, October 1955.

114 Titchmarsh, E. C., and Lapidus, L., "The Identification of Nonlinear Systems", Industrial and Engineering Chemistry, June 1967.

115 Towill, D. R., and Payne, P. A., "Frequency Domain Approach to Automatic Testing of Control Systems", The Radio and Electronic Engineer, Vol. 41, No. 2, 2, February 1971.

116 Townes, J. R., Dwyers, S. J., McLaren, R. W., and Zobrist, G. W., "Linear Techniques for Binary Fault Isolation", University of Missouri.

117 Ulrich, E. G., "The Evaluation of Digital Diagnosis Programs Through Digital Simulation", IEEE Conference, Publication No. 307, 1967.

118 Valstar, J. E., "Fault Isolation of Electronic Circuits Using Transfer Function Methods", Air Force Aero Propulsion Laboratory Wright, Patterson Air Force, Ohio, AD-608176, October 1964.

119 Valstar, J. E., "In Flight Dynamic Checkout", IEEE Transactions on Aerospace Support Conference, August 1963.

120 Verselova, G. P., and Gribanov, Yu, I., "The Optimal Sampling Step in the Calculation of Correlation Functions of Random Processes by Digital Computers", Automations Remote Control, Vol. 29, No. 12, pp. 1988-1994, 1968.

121 Wadsworth, R. B., and Thornton, B. M., "Industrial Systems Modeling", Industrial Engineering Department, Texas A and M University, ORSA 41st Meeting, 1972.

122 Walther, W., "Multiprocessor Self-Diagnosis, Surgery, and Recovery in Air Terminal Traffic Control", Transportation Systems Software Defense Systems Division, Sperry Rand.

123 Willer, D., "New Developments in System State Estimation and Anomaly Detection", ED.L.P. Hajden, SCI Final Report to BPA, October 1969.

124 Wilson, H., "Baker Sequences in System Identification Studies", Contro. Vol. 13, No. 127, pp. 48-51, 1969.

125 Wiston, G. C., Lombardo, J. J., and Tatge, R. B., "A New Diagnostic Technique for Hydraulic Systems", Control Engineering, May 1971.

126 Wolf, A. A., and Dietz, J. H., "A Statistical Theory for Parameter Identification in Physical Systems", Journal of the Franklin Institute, November 1962.

127 Woodrow, R. A., "Closed-Loop Dynamics from Normal Operating Records", Trans., of Instrument Technology, September 1958.

128 Wright, J. D., and Bacon, D. W., "Statistical Identification of the Behavior of a Dual Input Heat Exchanger Network", Internal Reports in Simulation, Optimization and Control, No. SOC-23, Faculty of Engineering, McMaster University, October 1973.

129 Young, P. G., "Process Parameter Estimation", Contr. Vol. 12, No. 125, pp. 931-936, 1968.

MODELING FAULTS IN DIGITAL LOGIC CIRCUITS

John P. Hayes

1 Introduction

Logic circuit design forms a major stage in the development of a digital system. The logic designer works with design components such as combinational networks and memory devices whose complexity may range from one or two logic gates to several thousand gates. Many different technologies are employed to construct digital systems. These technologies are characterized by rapid change, and a trend towards ever increasing complexity. The development of large-scale integrated circuit technology, for example, has made it possible to manufacture an entire computer on a single tiny chip of semiconductor material.

Digital circuits like all physical devices are subject to failure. A *fault* will be defined here as any change in a circuit that alters its function or behavior. While the introduction of more reliable technologies has tended to decrease the failure rate of individual logic devices, this has been offset to some extent by the higher complexity of the logic circuits being built today. In many applications such as the control of passenger-carrying vehicles, a fault in the logic circuitry can have disastrous consequences. Hence the problems of detecting and isolating faults in logic circuits are of great importance.

2 Physical versus Logical Fault Models

Given any physical fault mechanism in a logic circuit, it is always possible (at least in principle) to determine its effect on the logical behavior of the circuit. For example, Figure 1(a) shows a diode-transistor realization of a 3-input NAND gate G, which is represented symbolically in Figure 1(b). Two voltage levels V_H (high) and V_L (low) define the logic values 1 and 0

John P. Hayes is with the University of Southern California, Los Angeles, California 90007, in the Department of Electrical Engineering and Computer Science Program.

This research was supported by the National Science Foundation under Grant GK-37455 and the Office of Naval Research under Contract No. N00014-67-A-0269-0019.

respectively. The logical behavior of G is given by the truth table in Figure
1(c).

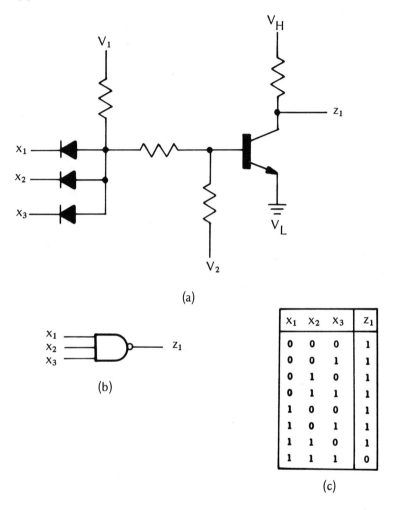

(a)

(b)

x_1	x_2	x_3	z_1
0	0	0	1
0	0	1	1
0	1	0	1
0	1	1	1
1	0	0	1
1	0	1	1
1	1	0	1
1	1	1	0

(c)

Figure 1. (a) Diode-transistor NAND gate G; (b) symbolic representation of G; (c) truth
table for G.

Suppose the following physical fault f_P occurs in G: input line x_1 is
accidentally connected (short circuited) to V_H. This could be expected to
have the effect shown in Figure 2(b) where the faulty output function z_1^* is
independent of the value of x_1.

We can readily define a logical fault f_L which mirrors precisely the effect of f_P. Disconnect input line x_1 from the fault-free gate G, and apply a constant logical 1 to the disconnected input as shown in Figure 2(a). This logical fault, usually referred to as "line x_1 stuck at logical 1" is equivalent to f_P. Note that while f_P may be an internal fault in G, the logical model assumes that G is fault-free and associates a fault with an interconnection line.

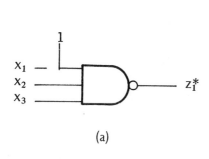

x_1	x_2	x_3	z_1^*
0	0	0	1
0	0	1	1
0	1	0	1
0	1	1	0
1	0	0	1
1	0	1	1
1	1	0	1
1	1	1	0

(a)

(b)

Figure 2. (a) The logical fault "input x_1 stuck at logical 1"; (b) truth table for the faulty output function z_1^*.

There are several advantages to using logical instead of physical fault models in digital fault analysis.

(1) Once we have a logical fault model that adequately reflects the physical failure modes of a circuit, fault analysis becomes a logical rather than a physical problem.

(2) It is possible to construct logical fault models that are applicable to many different technologies, in which case fault analysis becomes technology independent. This means that computer programs for fault simulation and test generation can be written that do not lose their usefulness with changes in technology.

(3) Using logical fault models it may be possible to derive tests for faults whose physical cause is unknown, or whose effect on circuit behavior is not completely understood.

3 General Remarks on Fault Types

Faults in digital circuits may be classified using the following criteria.

(1) The variation of the fault with respect to time.

(2) The effect of the fault on the logical function of the circuit components.

(3) The effect of the fault on the interconnection pattern (topology) of the circuit.

(4) The effect of the fault on the operating speed of the circuit.

(5) The number of distinct faults that may be present simultaneously.

It is usually assumed that faults are permanent, i.e., invariant with respect to time. Time-varying faults do occur, particularly intermittent faults, where the circuit moves in an apparently random fashion between the fault-free state and some fixed faulty state. An intermittent fault can only be detected if an appropriate test pattern is applied to the circuit under test during one of its faulty periods. The models that have been proposed for analyzing intermittent faults [1,10] require statistical data on the probability of fault occurrence; unfortunately this type of information is normally unavailable.

The types of logical faults in circuit components that need be modeled can often be restricted by very general physical arguments. Consider, for example, the 3-input NAND gate G of Figure 1 which realizes the function z_1 = $\overline{x_1 x_2 x_3}$.[1] It is unlikely that a fault can change z_1 to say, $\overline{x_1 + x_2 + x_3}$ (the NOR function), because this would require a major change in the switching threshold of the gate, i.e., in net input signal level at which the output signal changes value. Similarly, a change in gate function to $x_1 \oplus x_2 \oplus x_3$ is highly improbable as it would require a large increase in the number of threshold

---

[1]The following symbols will be used for the basic logical operations: juxtaposition (AND), + (OR), $\oplus$ (EXCLUSIVE OR) and overbar (NOT or COMPLEMENT).

levels. It can be assumed for most technologies that a fault will not change the basic gate type; a faulty NAND gate remains a NAND. This assumption enables us to restrict the possible faulty function for G to the following set of eight NAND functions (for simplicity, we will treat constant functions 0 and 1 as NAND's): $\langle 0, 1, \overline{x_1 x_2}, \overline{x_1 x_3}, \overline{x_2 x_3}, \overline{x_1}, \overline{x_2}, \overline{x_3} \rangle$. This is a very small subset of the $2^{2^3} = 256$ distinct switching functions of up to three variables.

Faults that change circuit topology include open circuits (the breaking of a connection) and short circuits (the establishment of a connection between two normally unconnected points). Open circuit faults are quite easy to model. The signal source end of an open connection has no further effect on the circuit. The sink end of the connection generally remains fixed at the logical 0 or 1 value as shown in Figure 3(a). Short circuit behavior is much more difficult to analyze, and fault models tend to be technology dependent. The signals on the sink ends of two short-circuited lines change from z_1 and z_2 to some functions $\varphi_1(z_1, z_2)$ and $\varphi_2(z_1, z_2)$ as indicated in Figure 3(b).

Changes in the signal propagation delay of the various gates and connections in a logic circuit can also cause erroneous behavior. For example, suppose the input pattern (x_1, x_2, x_3) applied to our 3-input NAND G is required to change at time t from $(1,1,0)$ to $(1,0,1)$. This should cause no change in the value of z_1 which should remain at 1. Suppose, however, that due to changes in signal delays, x_3 changes value Δt seconds before x_2, then $(1,1,1)$ will be applied to G for Δt seconds. If Δt is sufficiently long to overcome the inertia inherent in every physical device, a spurious 0 signal will appear at z_1.

The propagation delays of logic gates and connections are difficult to measure and can vary within wide tolerances. For this reason, conservative or worst-case design techniques are usually employed, as well as synchronization (clock) signals to compensate for minor variations in signal delays. Faults that result in asynchronous behavior are very difficult to analyze [17].

Finally we note that when $k \geqslant 1$ different parts of a circuit may be faulty at the same time, in which case we have a "multiple fault" of multiplicity k. If a particular fault model allows m independent single faults in a circuit, then there are $\sum_{i=1}^{k} \binom{m}{i}$ possible faults of multiplicity less than or equal to k. Frequently, fault analysis is based on a "single fault assumption" where $k = 1$. When k is large, measures must usually be taken to reduce the number of possible fault combinations to a manageable size.

Let M_1 and M_2 be two (possibly distinct) models for faults in some circuit N, and let F_1 and F_2 be sets of faults chosen from M_1 and M_2 respectively. F_1 is *equivalent* to F_2 if there is one-to-one correspondence between F_1 and F_2 such that $f_1 \in F$, corresponds to $f_2 \in F_2$ if and only if N realizes the same fault function with f_1 present as it does with f_2 present. If

F_1 and F_2 are equivalent then a test set T detects F_1 if and only if T also detects F_2. It is possible to isolate a fault to F_1 and F_2 only if F_1 and F_2 are non-equivalent.

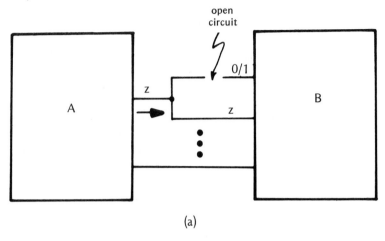

(a)

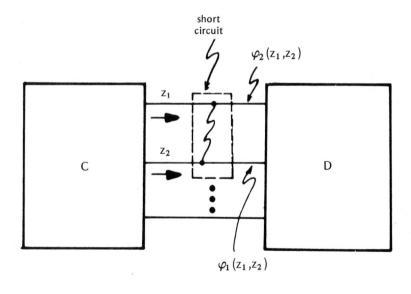

(b)

Figure 3. (a) Open circuit fault; (b) short circuit fault.

If every test set T that detects F_1 also detects F_2 then we will say that F_1 *strongly dominates* F_2. If some (but not necessarily every) minimal test set T for F_1 detects F_2, then F_1 *weakly dominates* F_2.

A knowledge of equivalence and dominance relations among faults is extremely useful. Given a class of faults F_2 in M_2 if we can determine an equivalent class of faults F_1 in M_1 which is simpler in some sense, e.g., better understood, then we can shift all our fault analysis from M_2 to M_1. If we have the weaker condition that F_1 dominates F_2, then we can transform the problem of generating a set of tests to detect F_1 into that of generating tests to detect F_2.

In the following sections, we will examine some representative logical fault models for digital circuits. Only permanent faults will be considered. In addition, in the next three sections we will restrict our attention to delay-free combinational circuits. However, all the fault models discussed in those sections are also applicable to sequential circuits.

4   The Single Stuck Line (SSL) Model

By far the most widely used logical fault model is the single stuck line (SSL) model. This model permits at most one logical connection in a circuit to be stuck at 0 (s-a-0) or stuck at 1 (s-a-1). Figure 2(a) shows a s-a-1 fault on the input line x_1 of G. The SSL model assumes that only one line is faulty at any time, and that all gates continue to function correctly in the presence of a fault.

The usefulness of the SSL model stems from the fact that it can be used to represent most of the physical faults that occur in logical circuits. As noted earlier, open circuits and certain kinds of short circuit faults immediately suggest this model. In addition, many internal failures in gates cause them to behave as if an input or output line of the gate were s-a-0 or s-a-1. The eight possible SSL faults in a 3-input NAND gate can transform its function to one of the following set

$$\langle 0, 1, \overline{x_1 x_2}, \overline{x_2 x_3}, \overline{x_1 x_3} \rangle$$

which includes five of the eight possible NAND functions noted in the preceding section. All SSL faults in this gate can be detected by applying the four tests $(0,1,1)$, $(1,0,1)$, $(1,1,0)$, $(1,1,1)$.

One of the primary tasks of digital fault analysis and test generation is relatively easy in the case of SSL faults. Most test generation methods employ a technique called *path sensitization* [4,6,15] which is illustrated in the circuit of Figure 4. Suppose we wish to find a test pattern for the fault "line y s-a-1." Clearly this test should place 0 on line y. This requires $(1,1)$ to be applied to G_4 as shown. A change in y from 0 to 1 due to the fault should cause a fault signal to propagate through the circuit to the primary output z_2, at which point it can be observed. This requires a signal path from y to z_2

that is sensitive to changes in the signal applied to y. Such a sensitized path is indicated by arrows in Figure 4. If the fault in question is present, all the logic values along this path change. The path from y to z_2 via G_3 is not sensitized. It is easily proven that the test $(x_1, x_2, x_3) = (1,1,1)$ is the only test for y s-a-1.

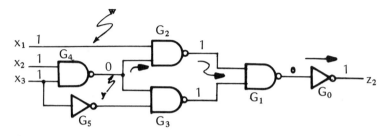

Figure 4. Example of a sensitized path

If a circuit contains p lines, the number of possible SSL faults is 2p, a relatively small number. It is certainly feasible to generate efficient test sets, i.e., with few unnecessary test patterns, for all SSL faults in any combinational circuit of practical size, and many algorithms have been developed for this purpose [4,6,15].

Not every potential physical fault is covered directly by the SSL model. In the circuit of Figure 4 it is conceivable that a short circuit fault could cause the output gate G_0 to be bypassed and change the circuit function from z_2 to its complement $\overline{z_2}$.

Theorem 1 [7]. No SSL fault in a non-redundant[2] circuit can change the output function from z to $\overline{z}$.

Proof: Suppose that some fault f_1 = "line y s-a-d" where d is 0 or 1 changes z to $\overline{z}$. This means that the response of the faulty circuit to every test pattern must be incorrect. There is some test t that checks for the SSL fault f_2 = "line y s-a-$\overline{d}$." To do so, t must apply the signal d to line y, and so cannot be affected by y s-a-d. This implies the output of the circuit is correct when t is applied and f_1 is present—a contradiction.■

---

[2] A non-redundant circuit is one from which no lines or gates can be removed without changing the output function. This is equivalent to saying that every stuck line fault is detectable. We will only consider circuits of this type.

Thus these are physical faults whose behavior cannot be modeled by SSL faults. In many cases such faults are dominated by SSL faults. For example, a fault f changing z to $\bar{z}$ is dominated by every SSL fault, since f can be detected by any test pattern. In view of the relative ease with which SSL tests can be generated, the ability of these tests to detect non-SSL faults is of considerable interest.

5 The Multiple Stuck Line (MSL) Model

An obvious extension of the SSL model is to multiple stuck line (MSL) faults by allowing more than one line to be s-a-0/1. This, however, permits an enormous jump in the number of possible faults in a p-line circuit from 2p to 3^P-1. It is this vast number of MSL faults that has caused them to be ignored until quite recently.

MSL faults are important for several reasons.

(1) Faults occurring during manufacture of a circuit frequently affect many parts of a circuit. The MSL model may therefore be more appropriate than the SSL model for acceptance testing after manufacture. Multiple faults are also encountered during field operation, particularly if the intervals between maintenance are long.

(2) Paradoxically, use of the MSL model can sometimes simplify SSL fault analysis. For example, in generating SSL tests for all-NAND circuits, s-a-0 faults can be ignored if one is willing to consider both single and multiple s-a-1 faults [7]. In a sequential circuit, a SSL fault may, over a period of time, propagate faulty signals to many different parts of the circuit. Multiple fault models may then be most appropriate for analyzing the subsequent behavior of the circuit.

(3) The class of single gate fault functions noted in the third section where the gate type does not change (NAND's remain NAND's etc.) is precisely the class of fault functions obtained by applying all MSL faults to the gate input lines.

It is possible for a test set to detect all SSL faults in a circuit, but fail to detect some MSL faults. This can be attributed to a phenomenon termed "fault masking." Consider again the circuit of Figure 4 and the test (1,1,1) which detects the fault f_1 = "line y s-a-1." If a second SSL fault f_2 = "line w s-a-0" is present, then no error indication appears at z_2, the primary output of the circuit, because the error signal from the site of f_1 is blocked (masked) at G_2 by f_2.

In general, an SSL test set fails to detect a multiple fault $(f_1, f_2, \ldots, f_k)$ if the detection of each component fault f_i is masked by the remaining faults [7]. The masking conditions for undetectability are relatively complex, and not frequently encountered in practice. This justifies the common assumption that SSL tests are adequate for MSL detection. However, it is usually not possible to guarantee that a given SSL test set covers all MSL faults without detailed analysis.

There are some interesting classes of circuits for which SSL tests are guaranteed to detect all MSL faults, i.e. the SSL faults dominate (either weakly or strongly) the MSL faults.

A restricted circuit [16] is a logic circuit satisfying the following two conditions.

(1) It contains no subcircuit with the same interconnection pattern as the 5-gate circuit of Figure 5(a).

(2) Only the primary input lines may fan out to two or more gates.

Figure 5(b) shows such a circuit. Restricted circuits include the important class of 2-level circuits.

Theorem 2 [16]. Every SSL test set for a restricted circuit is also an MSL test set.

A slightly weaker result exists for the class of fanout-free circuits. A *fanout-free circuit* is one in which every signal line is connected to at most one gate. This implies that there is only one path from every line to the primary output.

Theorem 3 [7,9]. For every fanout-free circuit there exists a minimal SSL test set which is also a MSL test set.

Not every SSL test set for a fanout-free circuit detects all multiple faults, but simple procedures are known for generating minimal SSL test sets that also detect all MSL faults [7]. Thus in restricted circuits the SSL faults strongly dominate the MSL faults, while in fanout-free circuits the SSL faults weakly dominate the MSL faults.

6 Short Circuit Faults

As noted earlier stuck line fault models may not be adequate for treating some types of short circuit failures. Recently a number of formal models for these faults have been proposed [5,11,13].

Consider the short circuit shown in Figure 3(b). In several important current technologies (primarily bipolar semiconductor circuits) the new functions $\varphi_1(z_1,z_2)$ and $\varphi_2(z_1,z_2)$ introduced by short circuit are such that

$\varphi_1 = \varphi_2 = \varphi$ and φ is either $z_1 z_2$ or $z_1 + z_2$. These fault functions are termed "wired AND" and "wired OR" respectively.

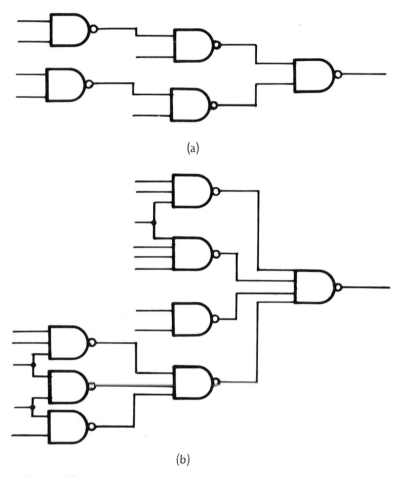

(a)

(b)

Figure 5. (a) Minimal unrestricted circuit; (b) example of a restricted circuit.

The detection of short circuit faults under the assumption that the short circuit creates a wired AND or OR is studied in [13] where they are termed *bridging faults*. The number of possible single bridging faults (involving just one pair of lines) in a p-line circuit is $\binom{p}{2} = p(p-1)/2$. It does not seem to be feasible to attempt to derive tests for all such faults. For example, a short circuit can introduce feedback, and so convert a combinational to a sequential circuit. Special types of bridging faults are amenable to analysis, however.

Practical experience indicates that most bridging faults are detected by a complete set of SSL tests. For certain bridging faults this can be guaranteed. An *input bridging fault* is a bridging fault that only involves the input lines of a single gate. It is assumed that there is no point of fanout between the site of the short circuit and the gate in question.

Theorem 4 [5, 13]. Every SSL test set detects all input bridging faults.

An interesting technique for modeling bridging faults has been proposed by Kaposi and Kaposi [11]. Suppose a short circuit between lines carrying z_1 and z_2 as in Figure 3(b) results in a wired OR. For fault analysis, replace these two lines by the circuit shown in Figure 6. Under fault-free conditions $\varphi_1(z_1,z_2)=z_1$ and $\varphi_2(z_1 z_2) = z_2$. If the fault f = "line x s-a-1" is present in this circuit, then both φ_1 and φ_2 change to $z_1 + z_2$. Hence the problem of detecting the bridging fault in the original circuit has been transformed into the problem of detecting the SSL fault f in the modified circuit. This modeling technique can be extended to a wide class of faults, its usefulness being limited by the complexity of the modified circuit.

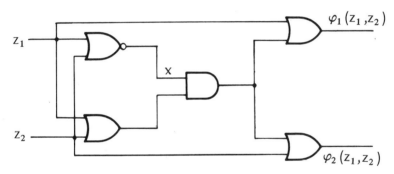

Figure 6. Model for wired OR short circuit fault.

Finally, we consider a type of short circuit failure called a *shorted diode fault* [5] which occurs in circuits constructed using diode-transistor technology of the type shown in Figure 1(a). Each input of the gate is connected to a diode. A shorted diode fault occurs when the input and output connections of one of these input diodes are short circuited. Thus a shorted diode fault is a short circuit that occurs inside a gate, and so is distinct from the bridging faults discussed previously.

A peculiarity of this type of fault is that the output of the faulty gate is unaffected, but the outputs of other gates connected to the faulty line are affected. Consider the AND circuit in Figure 7(a). Under normal conditions $z_1 = x_1 x_2$ and $z_2 = x_2 x_3$. Suppose that the diode in input line y_1 of G_1 is short circuited. If $x_1 = 1$, the correct signals appear at both outputs. If $x_1 = 0$, then $z_1 = 0$ and y_1 is forced to 0 independent of the value of x_2. Thus an incorrect response is obtained at z_2 for the test pattern $(x_1,x_2,x_3) = (0,1,1)$. The circuit in Figure 7(b) has the same behavior as the circuit in Figure 7(a) when y_1 contains a shorted diode. Clearly $z_1^* = z_1 = x_1 x_2$, but $z_2^* = x_1 x_2 x_3$. Thus the shorted diode fault has the same effect as moving the point of fanout from the input side to the output side of the faulty gate.

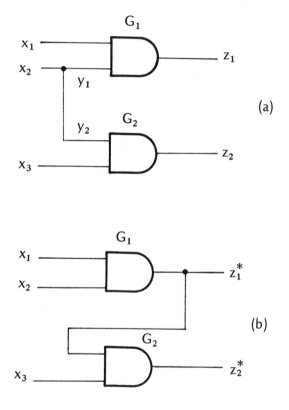

Figure 7. (a) Circuit with a shorted diode in line y_1; (b) logical fault model.

7 Pattern Sensitive Faults

We turn now to the problem of testing a particular class of sequential circuits called random access memories (RAM's). A RAM can be defined as an array $\langle C_0, C_1, \ldots, C_{r-1} \rangle$ of r independently addressable binary storage cells. These devices are constructed using magnetic cores or, more recently, using semiconductor technologies. While many RAM faults are determined by the technology used, there exists a significant class of faults called pattern sensitive faults (PSF's) which are common to various RAM technologies.

A PSF is said to be present when the result of a READ or WRITE operation performed on some cell C_i affects or is affected by the data (pattern) stored in the memory. The detection of this type of fault is of considerable interest to both manufacturers and users of RAM's. Unfortunately the physical fault modes that result in pattern sensitivity are not properly understood. For PSF analysis a RAM is treated as a proverbial "black box," and little use is made of circuit information.

Memories are usually tested for PSF's by applying standard easily generated test sequences of READ's and WRITE's to the memory. The following "walking 0/1" test [3] is typical. All memory cells are initially reset to the 0 state. Then a 1 is written into every cell in sequence beginning at C_0. After each WRITE the entire contents of the memory is checked for correctness by applying r READ's. The process is then repeated, this time writing 0's into successive cells. The length of this walking 0/1 test, i.e. the number of READ and WRITE operations needed, is approximately $2r^2 + 3r$.

Tests of the above type only check the RAM with a small fraction of the 2^r possible data patterns (states) present. It is not easy to classify the PSF's detected by these tests, and no attempt appears to have been made to do so. Since no fault model is postulated, such basic questions as how many faults are detected, or how near the test length is to being minimal, cannot be determined analytically.

A formal logical model for PSF's in RAM's has been proposed in [8]. An r-cell RAM M_r is considered to be an incompletely specified Mealy-type sequential machine with 2^r states, 3r input symbols (WRITE0, WRITE1 and READ for each of r cells) and two output symbols 0 and 1. Figure 8(a) shows the state table for M_2, where W_i, $\overline{W_i}$, R_i denote the WRITE1, WRITE0 and READ operations applied to cell C_i. Note that the output function z is unspecified (or a "don't care") for all WRITE operations.

An *unrestricted PSF* is defined to be any change in the specified values of the state table for M_r. Figure 8(b) shows the state table for M_2 with a PSF affecting W_0 and R_1 present. The problem of deriving a test for all unrestricted PSF's is the classic problem of finding a checking sequence[3] for M_r [6]. M_r has a number of properties that make it easy to construct an efficient checking sequence.

---

[3] A *checking sequence* for a sequential machine M is an input sequence whose response indicates whether or not the machine under test has the same behavior, i.e., the same state table, as M.

21

Input

State $y_0 y_1$ y		W_0	W_1	$\overline{W_0}$	$\overline{W_1}$	R_0	R_1
	00	10, -	01, -	00, -	00, -	00,0	00,0
	01	11, -	01, -	01, -	00, -	01,0	01,1
	10	10, -	11, -	00, -	10, -	10,1	10,0
	11	11, -	11, -	01, -	10, -	11,1	11,1

(a)

Input

State $y_0 y_1$		W_0	W_1	$\overline{W_0}$	$\overline{W_1}$	R_0	R_1
	00	00, -	01, -	00, -	00, -	00,0	00,0
	01	11, -	01, -	01, -	00, -	01,0	01,0
	10	11, -	11, -	00, -	10, -	10,1	10,1
	11	11, -	11, -	01, -	11, -	11,1	11,1

(b)

Figure 8. (a) State table for M_2; (b) state table for M_2 with a PSF present.

Theorem 5 [8]. M_r has a checking sequence S of length $(3r^2 + 2r)2^r$.

While S appears to be of near-minimal length, it is far too long to be of practical value. (A typical value of r is 1024). It is necessary to greatly restrict the number of possible PSF's if test sequences of practical length are to be achieved. One way of doing this is to assume that PSF's involving operations addressed to cell C_i are only influenced by patterns in a small fixed set of cells N_i called the neighborhood of C_i. PSF's of this type are called *local PSF's.*

The length of a test sequence to detect all local PSF's with respect to some set of neighborhoods is mainly a function of neighborhood size. The complexity of test generation is dependent on the degree of overlap among the neighborhoods. Consider the following situation. We wish to check some WRITE operation W_i addressed to cell C_i. After applying W_i to C_i we must read the contents of each cell C_j in N_i. However the response obtained from R_j may be influenced by the contents of N_i, and N_j may include cells outside N_i. Thus a complex chain of interactions can take place among the different neighborhoods. These difficulties can be largely avoided by making the following single fault assumption: only READ or WRITE operations involving a single memory cell can be faulty. Faults of this type are termed *single PSF's*.

Theorem 6 [8]. Let Q denote all single PSF's with respect to some neighborhood set $\langle N_1, N_2, \ldots, N_k \rangle$. Q can be detected by a sequence of the form $S_1 S_2 \ldots S_k$ where S_i is a checking sequence for N_i and $1 \leqslant i \leqslant k$.

Concluding Remarks

We have noted throughout this paper the relation between the SSL model and other fault models. This is significant in view of the great amount of effort that has been devoted to the analysis of SSL faults.

It is possible to treat many faults including stuck line faults and short circuit faults in a uniform manner. We can define a *"faulty line"* (FL) model[4] as follows.

(1) Let $k \geqslant 1$ lines in the original circuit be cut.

(2) Apply to the sink ends of the cut lines a set of k functions $\varphi_1, \varphi_2, \ldots, \varphi_k$.

If the functions $\varphi_1, \varphi_2, \ldots, \varphi_k$ in some FL fault F are all constants, then F is an MSL fault. (It is an SSL fault when $k = 1$.) If these k functions are AND or OR functions of the original signals on the cut lines, then F models a bridging fault. The shorted diode fault in Figure 7 can also be covered by cutting line y_2 and applying the function $x_1 x_2$ to the cut line.

At any point in time in a circuit with an FL fault of multiplicity k, a pattern of k 0's and 1's is being applied to the k faulty lines. Hence an FL fault F behaves like a time-varying MSL fault. In many circuits this time variation is relatively slight as only a small part of the circuit is active at any time. Hence over a short period F may behave like a permanent MSL fault. As noted in the section on the MSL model, in most circuits the class of all SSL

---

[4]This model is general enough to cover *any* possible permanent fault in a combinational circuit. If a fault producing output function z^* is to be modeled, cut the output line of the original circuit and apply $\varphi_1 = z^*$ to the cut line.

faults dominates the MSL faults. Hence the foregoing argument suggests that in many circuits the SSL faults are likely to dominate both MSL and short circuit faults, a fact that has been confirmed experimentally.

Although the fault models we have examined appear adequate to treat most of the faults encountered in current technologies, there is no guarantee that this will be the case in the future. New logic technologies such as magnetic bubble logic [14] continue to emerge whose physical characteristics and fault modes differ considerably from earlier technologies. Furthermore, existing fault models have many inadequacies, particularly in the area of sequential circuits. Hence it is likely that fault modeling will continue to be a problem facing the logic circuit designer.

REFERENCES

1 M. A. Breuer, "Testing for intermittent faults in digital circuits," *IEEE Trans. Computers*, vol. C-22, pp. 241-246, March 1973.

2 M. A. Breuer, "Modeling circuits for test generation," *Digest of Papers, Fourth Internat. Symp. on Fault Tolerant Computing*, pp. 3.13-3.18, June 1974.

3 J. R. Brown, "Pattern sensitivity in MOS memories," in *Testing to integrate semiconductor memories into computer mainframes*, Digest of Symposium, Cherry Hill, New Jersey, pp. 33-36, October 1972 (available from IEEE).

4 H. Y. Chang, E. G. Manning and G. Metze, *Fault diagnosis of digital systems*, Wiley, New York, 1970.

5 A. D. Friedman, "Diagnosis of short-circuit failures in combinational circuits," *IEEE Trans. Computers*, vol. C-23, pp. 746-752, July 1974.

6 A. D. Friedman and P. R. Menon, *Fault detection in digital circuits*, Prentice-Hall, Englewood Cliffs, New Jersey, 1971

7 J. P. Hayes, "A NAND model for fault diagnosis in combinational logic networks," *IEEE Trans. Computers*, vol. C-20, pp. 1496-1506, December 1971.

8 J. P. Hayes, "Detection of pattern sensitive faults in random access memories," *IEEE Trans. Computers*, vol C-24, pp. 150-157, February 1975.

9 K. Kajitani, Y. Tezuka, and Y. Kasahara, "Diagnosis of multiple faults in combinational circuits," *Electronics & Communications in Japan*, vol. 52-C, pp. 123-131, 1969.

10 S. Kamal and C. V. Page, "Intermittent faults: a model and a detection procedure," *IEEE Trans. Computers*, vol. C-23, pp. 713-719, July 1974.

11 J. F. Kaposi and A. A. Kaposi, "Testing switching networks for short circuit faults," *Electronics Letters*, vol. 8, pp. 586-587, November 1972.

12 E. J. McCluskey & F. W. Clegg, "Fault equivalence in combinational logic networks," *IEEE Trans. Computers*, vol. C-20, pp. 1286-1292, November 1971.

13 K. C. Y. Mei, "Bridging and stuck-at faults," *IEEE Trans. Computers*, vol. C-23, pp. 720-727, July 1974.

14 R. C. Minnick et al, "Magnetic bubble computer systems," *Proc. 1972 Fall Joint Computer Conf.*, pp. 1279-1298, December 1972.

15 J. P. Roth, "Diagnosis of automata failures: a calculus and a method," *IBM Journ. Research & Development*, vol. 10, pp. 278-291, July 1966.

16 D. R. Schertz and G. Metze, "A new representation for faults in combinational digital circuits," *IEEE Trans. Computers*, vol. C-21, pp. 858-866, August 1972.

17 S. H. Unger, *Asynchronous sequential switching circuits*, Wiley-Interscience, New York, 1969.

COMPUTER-AIDED FAULT ANALYSIS—
TODAY, TOMORROW, OR NEVER?

J. R. Greenbaum

The title indicates a slightly schizophrenic attitude about ever being able to achieve a totally satisfactory capability to test electronic circuits and systems automatically with the computer programs, procedures, and philosophies presently used. I think that part of the problem is that today we sit on a patchwork of computer programs that, like Topsy, "just growed." Most of the programs were developed independently; first to assist the circuit and system design engineer, then to assist the draftsman, and finally to assist in testing the resultant product. What appears to be necessary is an integrated group of programs to reflect the optimum available from each area. An indication of effort that has gone into the test point generation type of programming is shown in Figure I, which is a copy of a page from a Navy Department document [1]. As can be seen, even in early 1973, sixteen different programs had been developed and advertised. There is no question that there are an equal number about which we have not heard. Developing, testing, and placing this number of programs into operation obviously represents a considerable expenditure of both time and money by industry and thus reflects the importance given by industry to the solution of automated testing problems.

For those companies with a complete set of computer programs, i.e., for design, drafting, layout, and wiring, as well as for testing, an additional large investment of time and money has gone into marrying these several areas of computer assisted effort. The results, although very usable and much superior to anything that has been developed as a single, integrated program, are, in my opinion, the absolute minimum. This is despite the fact that the independent requirements for the different areas and the independent growth in these areas are probably two of the major reasons for the limited capabilities that exist today.

The purpose of this paper is to try to present the point of view of the program user; that is, the person who applies these programs to equipment that must be delivered to our customers. (Our customers are the several

J. R. Greenbaum is with the Re-entry and Environmental Systems Division of the General Electric Co., Phila, Pa.

branches of the military.) Therefore, this paper describes how we in the defense electronics department at General Electric view the problem both in an immediate and on a long term basis. Then, since I consider General Electric to be reasonably representative of industry, I will try to indicate where I feel industry should be heading. Finally, I will make some suggestions on how to get there from here.

DIGITAL ATPG	DEVELOPED BY	COMPANIES USING ATPG
1 D-LASAR	DIGITEST INC.	DIGITEST INC. LOCKHEED GEORGIA CO.
2 FAIRSIM II/FAIRGEN	FAIRCHILD	FAIRCHILD
3 TESTGEN/COMTEST	WESTINGHOUSE	WESTINGHOUSE AAI OTHERS
4 FAULTS II	GENERAL DYNAMICS	GENERAL DYNAMICS
5 SALT	IBM	IBM
6 LASAR II	LTV	LTV PRD
7 TGEN	RCA	RCA
8 TESTAID	TELPAR	TELPAR HP SEVERAL OTHERS
9 TASC	PACIFIC APPLIED SYSTEMS	PAS OTHERS
10 FLASH	MICRO INC.	MICRO SPERRY RAND GENERAL RADIO
11 SATGEN	HUGHES AIRCRAFT	HUGHES
12 ATVG	GE	GE
13 FAS/SDAP	HONEYWELL	HONEYWELL
14 LOGOS	GRUMMAN	GRUMMAN
15 TGAS	NAVY	NAVY
16 SEQUENTIAL ANALYSER	BELL LABS	BELL LABS

Figure 1. List of test point generation programs.

In order to appreciate some of the "hows" and "whys", it is necessary to understand how we undertake a new design effort. In general, any new equipment task can be divided into four areas, as shown in Figure 2. Once a system concept has been determined, system implementation is normally achieved by reducing the design concept to building blocks. In engineering design, the building block functions are subdivided further until they can be apportioned to individual engineers who are responsible for implementation. Then when the design progresses sufficiently, the documentation effort is started. After the product is fabricated, testing is initiated to verify that the end product performs as required.

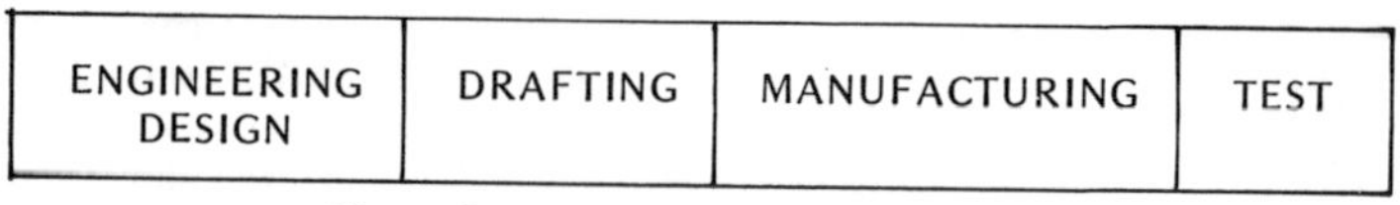

ENGINEERING DESIGN	DRAFTING	MANUFACTURING	TEST

Figure 2.   New equipment task areas.

Today the primary area of concern in automated testing is the world of digital systems; therefore, this paper will be limited to this area. However, let us recognize that the concern for computer testing of analog circuits has existed for some time also. For example, an excellent article that appeared in the November 15, 1972, issue of E.D.N. describes how to evaluate circuit performance by simulating various modes of transistor failure using Computer-Aided Design (CAD) circuit analysis programs.   To assist the digital circuit/system design engineer during the design stage, we at General Electric use a computer program known as Logic and Timing Simulation (LATS). Programs of this type allow the designer to detect and analyze two classes of error: logic and timing. Since this is done while the design is still in the conceptual stage, the analyses are rapid, economical, and efficient. The program accepts logic networks composed of elements of any type, so that if device parameter information is available, it can be used on the small chip (NAND gate for example) or on a large functional basis, as desired. The device performance characteristics are stored in a program library which is accessed as needed. In the process of designing the digital function, the program simulates a considerable number of input-output conditions. These obviously can be reflected into the first test program.

At the fundamental level, the result of a LATS type of analysis is an engineering generated, hand-drawn logic diagram which is sent to the drafting function (Figure 3). We use a program called ALI (Automatic Logic Implementation) which results in a physically partitioned series of units that are correctly identified and documented, a parts listing, a running list, and a properly drawn schematic. Also, information in the form of lists and tapes is provided for production layout and the automated testing program. Figure 4 shows a partitioned section of the Figure 3 input; Figure 5 shows a typical computer description of the wire running list.

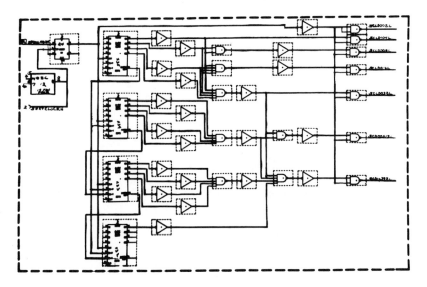

Figure 3. Designer's worksheet.

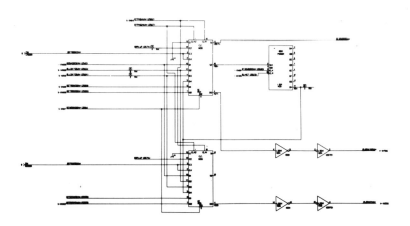

Figure 4. ALI output.

R E V	SIGNAL IDENT	SEQ NO.	FROM (RCP)	TO (RCP)	FUNCTIONAL NAME	NOTES
	- HALL	001	A9111	A0549		
	-1.5V	001	A9114	A9407		
		002	A9407	A0612		
		003	A9407	A0529		
	-12V	001	A9103	A0545		
		002	A9540	A0632		
		003	A9396	A0545		
	+ HALL	001	A9110	A0951		
	+12V	001	A9538	A0606		
		002	A9126	A0534		
	+18V	001	A9120	A0127		
	+5.9V	001	A9113	A0551		
		003	A9520	A0625		
	+5V	001	A9104	A0302		
		002	A9102	A0614		
	ACS OUT A1	001	A0227	A0332		
	ACS OUT A10	001	A9236	A0339		
	ACS OUT A11	001	A9237	A0342		
	ACS OUT A12	001	A9238	A0341		
	ACS OUT A13	001	A9239	A0344		
	ACS OUT A14	001	A9240	A0343		
	ACS OUT A15	001	A9241	A0346		
	ACS OUT A16	002	A9242	A0345		
	ACS OUT A17	001	A9243	A0340		
	ACS OUT A18	001	A9244	A0347		
	ACS OUT A19	001	A9245	A0350		
	ACS OUT A2	001	A0228	A0331		
	ACS OUT A20	001	A9246	A0349		
	ACS OUT A21	001	A9247	A0352		
	ACS OUT A22	001	A9248	A0351		
	ACS OUT 23	001	A0249	A0354		
	ACS OUT A24	001	A9250	A0353		
	ACS OUT A3	001	A9229	A0334		
	ACS OUT A4	001	A9230	A0333		
	ACS OUT A5	001	A9231	A0336		
	ACS OUT A6	001	A9232	A0339		
	ACS OUT A7	001	A9233	A0338		
	ACS OUT A8	001	A9234	A0337		
	ACS OUT A9	001	A9235	A0340		
	ACS OUT B1	001	A9203	A0308		
	ACS OUT B10	001	A9212	A0315		
	ACS OUT B11	001	A9213	A0318		
	ACS OUT B12	001	A9214	A0317		

Figure 5. ALI generated running list.

To test for proper performance, we use such computer programs as ATVG (Automatic Test Vector Generation), D-Laser, TGAS, or FAS (Fault Analysis Simulation). Each of these programs has its proponents, since each has its advantages. Simulation programs such as FAS, provide the following types of information:

1. Verification of error free test vectors for functional static testing. This excludes precise measurement of voltage, current, resistance, or time. The tests are static in the sense that the responses are all stable at the time they are sampled; that is, they are constant at either a logic 0 or a logic 1 when measured.

2. Automatic test station capability. The program provides a paper or magnetic tape which on a system or subsystem basis allows the test station to detect a bad unit in 1 or 2 minutes versus possibly hours if a manual test were to be conducted on a large subsystem.

3. Diagnostic information which should reduce the time it takes to isolate faults from hours to minutes.

   a. The response of every terminal pin of every' integrated

circuit (IC), and, of course, for every accessible output pin is given for every input test vector.

b.   The IC's suspected of causing the abnormal output are specified.

4.   A quantitative measure of the efficiency of the tests is explicitly provided; i.e., the results of all tests are printed upon test completion.

5.   Formal documentation is available from the computer to form the bulk of an official test specification document.

One of the main reasons for the use of several of these programs in the company is the lack of a uniform library of device models for the many integrated circuits that need to be simulated so that proper circuit behavior can be predicted. A standardized library of over 200 LSI devices and 50 NAFI-SHP modules has recently been established to be used by both the circuit design programs and the fault analysis/testing programs. This is expected to result in a commonality of applicable computer programs to take us quickly from design through test.

However, even this marriage of programs and library standardization is visualized as an answer only to the short term problem and still begs the issue for the longer term approach. A few of the obvious long term questions that must be resolved are:

1.   When new equipment is being designed, what should be done to improve automatic testability?

2.   What can we do to design and/or partition new equipment automatically to result in large improvements in our ability to test it and readily define faults?

3.   To what degree should realiability be factored into fault analysis plans? Obviously, high failure rate items will be a problem.

4.   How do we evaluate computer program improvements which result in more rapid fault analysis when the improvements might better be added to the ability to handle larger or more complex and sophisticated circuits?

5.   Should we be putting our efforts into trying to define a better way to design equipment that includes its failure analysis/ testability, or are other approaches more desirable?

6.   What priority should be assigned to the solution of immediate versus long term problems?

These few questions seem almost to generate an equivalent number of additional questions. Therefore, I would like to suggest two major areas of investigation that should yield maximum results and answer most of the questions that come to mind. The first to accommodate the more immediate problems; the second, the longer term considerations.

The first area consists of guidelines that must be considered during the design stage to make a design cost-effectively-testable. I believe that following these guidelines will bring test costs down to an acceptable level without in any way impairing system performance. In addition, the guidelines should simplify the test process to such a degree that the need for designers to assist in troubleshooting is minimized. This implies that with on-site test programs, less technically trained personnel can readily troubleshoot and maintain complex equipment. For the military this implies a minimum of equipment downtime resulting from the minimization of MTTR (Mean Time To Repair).

Figure 6 shows the eight elements of concern. Note that they apply to technical elements that experience has shown to be major problem areas in automatic testing. They are somewhat obvious, but are the result of considerable experience.

1. PHYSICALLY PARTITION THE HARDWARE TO FORM READILY TESTABLE PARTITIONS.

2. PROVIDE RESET (OR SET) LINES FOR INITIALIZING THE NET-WORK.

3. PROVIDE "BREAKS" IN CASCADED COUNTER CARRY (TERMI-NAL COUNT) LINES OR PROVIDE BUILT-IN MULTIPLEXER CIR-CUITS.

4. MAKE AVAILABLE CONNECTIONS WHICH CONTROL THE STATES OF MEMORY ELEMENTS.

5. PACKAGE ROM, RAM, AND LARGE SHIFT REGISTER LOGIC TO FORM SEPARATE TESTABLE PARTITIONS.

6. PROVIDE ADEQUATE I/O PINS AND CONSIDER USING SPARE IC LOCATIONS TO OBTAIN ADDITIONAL ACCESSIBLE TEST PINS.

7. INHIBIT FREE RUNNING CLOCKS AND ALLOW FOR INPUT TEST CLOCKS.

8. PERFORM LOGIC AND TIMING SIMULATION (LATS) AND UTI-LIZE LATS VECTORS AS STARTING POINT FOR SUBASSEMBLY TEST.

Figure 6. Guidelines.

Taking the items one at a time, the following are the reasons for each suggestion:

1.   Physically Partition the Hardware to Form Readily Testable Partitions; Take Advantage of All Unused Connector Pins.

After the design function is satisfied, the first priority for use of unused connector pins should be to form testable partitions which may be called subsets of the total network having accessible I/O pins so that they may be independently tested. If large areas of logic (i.e., more than 40 IC's) can be "broken up" by actually opening certain lines at critical points, testability will be greatly enhanced. The critical points may be as few as three or four signals between subfunctions of a logical system function. The lines are broken by creating an accessible output and an accessible input in each initially shorted line. Figure 7 illustrates this guideline which provides the following benefits:

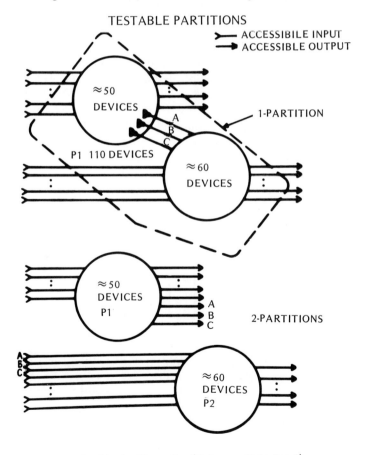

Figure 7. Partitioning illustration (Reference Guideline 1).

a. A complex network becomes two less complex networks.

b. The time it takes to generate test vectors is greatly reduced (factors of 10 to 1 can be realized).

c. The impact of a design change on cost and schedule time to update test vectors is minimized. Typically one or more changes affects only one testable partition; only that partition requires rework.

d. Fault resolution areas become smaller and troubleshooting time is minimized at the test station.

2. Provide Reset (or Set) Lines for Initializing the Network.

Before a network can be tested, it must be placed in a known state. Testability is restricted in system designs having memory elements embedded in logic such that their control inputs are in complex feedback loops.

The problem is solved by providing one or more accessible inputs for the sole purpose of placing memory elements into a known state. This guideline is demonstrated in Figure 8 for the hypothetical case of five devices which can be initialized with a logic 0 on their asynchronous inputs. Input pins A12 and B14 are grounded during normal system operation and for test purposes would have a logic 1 applied to initialize the five devices.

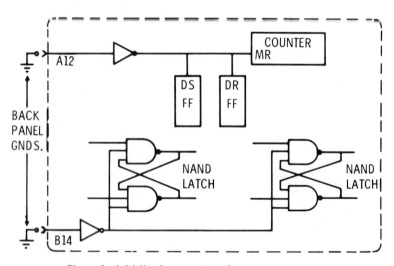

Figure 8. Initialization capability (Reference Guideline 2).

The following benefits are realized using this guideline:

a.   Fault isolation time at the test set is considerably reduced.

b.   Test vector generation time is considerably reduced.

c.   Computer simulation time is reduced by decreasing the number of excitations needed.

    (1)   Some subnetworks requiring initialization may be common to several testable partition networks.

    (2)   The initial state may be required for many different test vector sets for the same testable partition. Each time a set of test vectors is run and found to be inadequate, additional vectors must be added and the network re-initialized.

d.   It avoids manual intervention in -

    (1)   the computer/test set interface

    (2)   the test set process

3.   In a Manner Analogous to Guideline 3, Provide "Breaks" in Cascaded Counter Carry (Terminal Count) Lines or Provide Built-In Multiplexer Circuits.

If we are to gain the benefits of computer simulation, we must also live with its constraints. Surprisingly, the key constraint is not any lack of program capability but rather the prohibitive cost of simulating certain hardware actions. This guideline allows for economical simulation. Consider the network shown in Figure 8. It has a 16-bit cascaded counter made up of four IC counters. The first problem that may exist here is one of initialization discussed earlier. The second problem is the fact that computer cost is proportional to the number of excitations required. To exercise pin E-16 requires $2^{16} = 65,536$ excitations (3 per clock; 0, 1, 0). A fault analysis run for a very large network can cost about \$150 for 65 excitation conditions. Making a conservative assumption that the counter is in a testable partition network costing 1/10 (or \$15 per 65 excitations) for 1000 tests of 65 excitations each, we have the prohibitive result that

$$\text{Computer cost} = \frac{\$15.00}{65 \text{ Excitations}} \times 3 \times 6500 \text{ excitations} = \$45,000.00$$

Although this assumes every state of the counter is obtained, which is not generally required, it remains a good yardstick for the following reasons:

a.   Generally a large subset of the counter states is required.

b.   It is necessary to obtain many of the required states more than once. (The same counter state must be repeated for different network conditions.)

The approach, shown on Figure 9, of providing breaks in the four output lines was used, and to date no ill effects have been known to result.

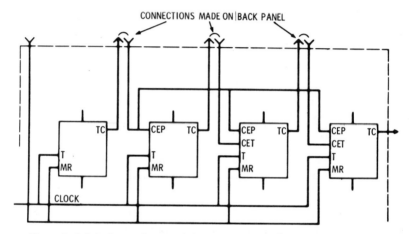

Figure 9. A Solution to the cascaded counter problem (Reference Guideline 3).

4.   Make Available Connections Which Control the State of Memory Elements.

This guideline is a generalization of guideline 2 (the Initialization Problem) and guideline 3 (the Cascaded Counter Problem). Adherence to this guideline will resolve most of the initialization problems. The general problem considered here involves situations where memory elements are embedded in feedback loops so that hours of analysis normally are required to obtain each of the many states needed throughout the network to test it effectively. Unfortunately, this dilemma may not be appreciated if one has not experienced the problem. It is one of the reasons that a complex System Replaceable Assembly (SRA) has been known to require as much as 3 man-months of manual Test Vector Generation Time. The problem exists for any method of testing, it is not unique to CAD; it is an outgrowth of the new generation hardware.

Testability is greatly enhanced by gaining access to the asynchronous inputs of the memory devices. This applies whether or not the asynchronous inputs are used as part of the design function. The guideline is demonstrated in Figure 10 . Points A and B were originally shorted; now they become shorted via a back panel connection; A18 becomes a control input for test purposes. Key benefits resulting from this approach are:

a.   Test vector generation time can realize reductions of 50 percent of more.

b.   Computer simulation and test time will be reduced.

c.   The impact of design changes on the time it takes to update test vectors is minimized.

d.   Fault isolation time at the test station is minimized.

5.   Package ROM, RAM, and Large Shift Register Logic to Form Separate Testable Partitions.

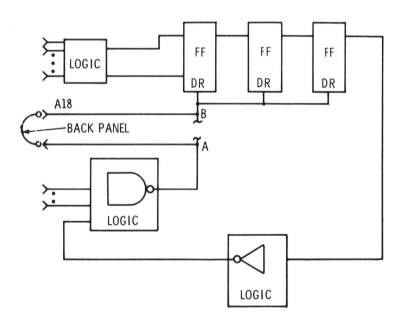

Figure 10. Controlling the states of memory elements (Reference Guideline 4).

The advent of IC chips containing 1024 and 2048 bits of information necessitates large computer memory requirements for the simulation program.

The simulation technique itself is not the problem but computer memory requirements for program simulation may become very high. This occurs when 40 or more of these IC's are present in the network. In fault mode simulation, 1 or 2 of these IC's can introduce thousands of fault modes. These factors could make computer simulation cost prohibitive.

The problem is aggravated even further when these IC's are embedded in logic; their address and control lines are excited by accessible input pins only through relatively large sections of logic and their outputs are observed on accessible outputs only through more logic.

Thus, it appears that the most economic approach to testing large ROMS, RAMS, and Shift Registers is to package them in separate testable partitions. Since their logic is straightforward, it suffices to detect faults through an exhaustive function-oriented approach with no diagnostics necessary. Computer simulation should not be necessary, except possibly on a fault free basis (no diagnostics), and then, simply to provide documentation compatible with other subassemblies.

Packaging these functions as separate testable partitions, by definition, means they should be independent entities with their inputs and outputs easily accessible. This was done for the ROMS in recently designed equipment and worked out well.

6. Provide Adequate I/O pins and Consider Using Spare IC Locations to Obtain Additional Accessible Test Pins.

Adherence to this guideline should overcome some of the limitations of the previous guidelines (solves the pin limited problem) if considered during the design and layout effort.

If wire-wrap type packaging or dual in-line with socket packing is used, then a dummy dual in-line package might be used in a spare location that could be removed during test and provide the necessary shorts on broken lines when replaced for system operational mode.

For a number of General Electric products, the System Replaceable Assembly (SRA) presently consists of two multilayer boards of 8 to 10 layers each. One board may have as many as 135 to 168 IC'S. Present configurations consist of a 300-pin bottom connector (accessible as input or output pins) with 140 terminal board (TB) crossover connections (accessible output pins only) on top between boards. This packaging configuration inherently results in large complex networks lacking testable partitions and accessible control inputs. It appears that a replaceable TB connector* should be used to provide the following:

a.   Automatically halve the complexity of the Unit Under Test (UUT). (The UUT becomes the board instead of the SRA.)

b.   Make TB connections accessible as inputs as well as outputs (presently restricted as output only).

This would provide much needed control inputs (automatically); and allow for further partitioning of the individual boards.

7.   Inhibit Free Running Clocks and Allow for Input Test Clocks.

Problems exist at the subassembly test level if a clock source is not accessible. The clock inputs to logic must be made accessible to synchronize them with other test data inputs. In addition the clock rates used for static subassembly testing are usually at a much slower rate than the operational system clock rates.

Hence the following points should be considered whenever free running clocks are present on the Unit Under Test (UUT).

a.   Can the clock be inhibited externally to allow for external application of a text clock?

b.   If the free running clock cannot be inhibited externally, then the UUT should be redesigned to allow for external test clock capability. Two alternative methods are shown in Figure 11.

8.   Perform Logic and Timing Simulation (LATS) and Utilize the LATS Vectors as Starting Point for Subassembly Test.

Most designs require LATS (or some equivalent computer program) for design verification. Because the objective in LATS is not directly concerned with verification of the excitation vectors for the purpose of fault detection, excitations used in LATS will generally be incomplete for test purposes. However, LATS

---

*This replaceable connector would provide 280 test points accessible as inputs or outputs when removed and 140 shorts between boards when replaced for system operation.

vectors may serve as a starting set of test vectors, and thus save considerable time in the generation of a complete set of test vectors; consequently, they should be documented with this in mind.

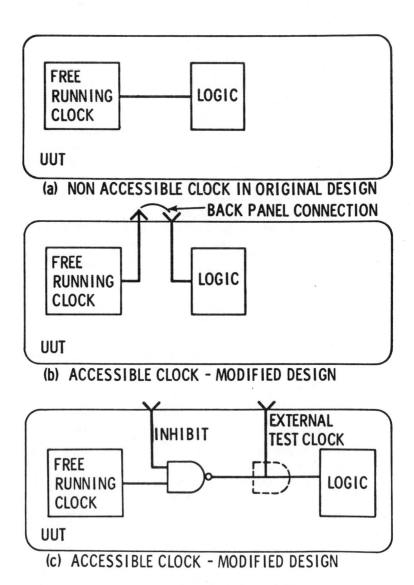

Figure 11. Solution to free running clock design (Reference Guideline 7).

Thus, we see that as we reflect the thesis of testability back to the original design, we can eliminate the unintentional lack of testability hidden in new designs; we can make the new equipment more reliable, and we can more readily predict its behavior through computer simulation.

That describes the first area, i.e., guidelines for those areas where we continue down the same path we are presently following and try to make it more cost effectively testable.

The second area, which we are just beginning to study in depth, requires a significant change in philosophy by the designer. Instead of following his present procedures of hand and/or computer analysis, and adapting the computer for automatic test procedures, the designer will have to adopt to what I have defined as a "design for testability" philosophy. In light of the advances in our capability to package more and more complex circuits on smaller chips, I feel that it will be more cost effective to add whatever logic is required to make designs self-diagnosing at the replaceable module level. This approach in essence eliminates the need to conduct extensive testing at all levels of circuit/function/system performance. It requires only that selected signals be monitored periodically and appropriate flags established if faulty performance is determined; the status of these flags would be continuously monitored. Whether this new approach will live up to its advanced billing remains to be seen, but if it does, a major obstacle in the automatic evaluation of the performance of large, complex, sophisticated electronic systems may be eliminated.

These are the two major areas that we in industry, who consider ourselves to be the program users, are presently pursuing. It is our expectation that this effort will provide reliable hardware that is easily designed, tested, and maintained at an economical cost.

Acknowledgments:

Some of the information included in this paper was generated by Mr. V. DeBuono of the General Electric Company, A.E.D., Utica, New York.

REFERENCE

1. *S-3A Automatic Test Equipment and Automatic Test Program Generation for Avionic Shop Replaceable Assemblies.* U.S. Navy Air Engineering Center, Philadelphia, Pa., Ground Support Equipment D.-GSED-MISC-107, February 1973.

RELIABLE DESIGN OF SOFTWARE

John F. Meyer

Introduction

Beginning with the pioneering work of von Neumann in 1952 [1], a variety of formal system models have been used to study questions concerning the reliability of computing systems. Questions that have been so considered range from the design of fault-tolerant switching networks to the verification of computer programs. In general, the ability to rely on a computing system (whether the system be implemented by hardware, software or both) is determined by the nature of potential deviations from the "desired" or "intended" system. The causes of such deviations are usually referred to as "faults" and, accordingly, the desired system is alternatively referred to as "fault-free."

In modeling systems with faults, there are basically two types of faults to be considered:

I.   (<u>Life</u> faults) : Faults which occur during the utilization of the system.

II. (<u>Birth</u> faults): Faults which occur during the design and implementation of the system.

Accordingly, models of systems with life faults are used to investigate the reliability of system utilization under the assumption that the system is fault-free when put into use. Systems with birth faults are used to investigate the reliability of system design and implementation, beginning with a specification of the desired fault-free implementation. This paper is primarily concerned with the latter problem where the systems in question are software systems. Beginning with the general notion of a "system with faults," the intent of the paper is to distinguish, via the formalism of the general model, specific areas of investigation pertaining to the reliable design of software. This is followed by a brief survey of research that has been performed to date in each area.

Systems with Faults

As first introduced in [2], the general concept of a "system with faults" provides a uniform way of viewing both life faults and birth faults.

John F. Meyer is a Professor in the Departments of Electrical and Computer Engineering and Computer Science at the University of Michigan, Ann Arbor, Michigan.

Informally, a system with faults is a system, along with a set of potential faults of the system, and a description of what happens to the original system as the result of each fault. These are the basic ingredients that are represented in the general model and comprise things one would like to know about the system even though they may not be known explicitly in a given application. More precisely, the original system and the systems resulting from faults are represented by members of one of two prescribed classes of (formal) systems, a "specification" class for the original system and a "realization" class for the resulting systems. More precisely, we say that a triple (S, R, ρ) is a (system) representation scheme if

(i) S is a class of systems, the specification class.
(ii) R is a class of systems, the realization class.
(iii) $\rho : R \to S$ where, if $R \in R$, R realizes $\rho(R)$.

By a class of systems, in this context, we mean a class of formal objects that are mathematical representations of real systems at some desired level of abstraction. Some examples of a representation scheme (S, R, ρ) are the following:

Example 1

S = All sequential machines
R = All sequential switching networks
$\rho : R \to S$
where $\rho(N)$ = The sequential machine that represents network N.

Example 2

S = All partial recursive functions
R = All PL/I programs with integer (fixed) variables
$\rho : R \to S$
where $\rho(P)$ = The partial recursive function realized when program P
                 is executed

Relative to the concept of a representation scheme, the general notion of a "system with faults" can be defined as follows: A system with faults in a representation scheme (S, R, ρ) is a structure (S, F, φ) where

(i) $S \in S$.
(ii)   F is a set, the faults of S.
(iii) $\varphi : F \to R$ such that, for some $f_0 \in F$, $\rho(\varphi(f_0)) = S$.

The intended interpretation of a system with faults is a (potentially) unreliable system where S represents the original specification of the system, F represents the set of potential faults that can occur in the process of realizing S, and, for a given fault $f \in F$, $\varphi(f)$ is the realization that results from the occurence of f. $\varphi(f)$ will alternatively be denoted S^f and referred to as the result of f. The fault f_0, whose existence is guaranteed by condition iii), represents the absence of an actual fault since the result of f_0 realizes S. More generally, if $f \in F$ and $\rho(S^f) = S$, we say that f is benign and S^f is fault-free. In these terms, condition iii) guarantees that the realization class will contain a fault-free realization of specification S.

To illustrate the generality of this concept, consider the following two examples. The first is a model of a "hardware" system with life faults; the second is a model of a "software" system with birth faults.

Example 3

Consider the representation scheme (S, R, ρ)
where
$S = R = \langle S | S$ is a network of sequential switching systems$\rangle$
and ρ is the identity function on R. Let $S \in S$

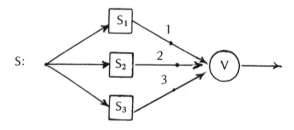

which employs triple modular redundancy, that is, $S_1 = S_2 = S_3$ and V is a voter. Suppose further that the potential faults are combinations of stuck-at-0 and stuck-at-1 faults at nodes 1, 2 and 3. Then (S,F,φ) is a system with faults where

S is as above

$F = \langle (a_1, a_2, a_3) \mid a_i \in \langle 0,1,x \rangle\rangle$

$$\text{where } a_i = \begin{cases} 0 \text{ if node i stuck-at-0} \\ 1 \text{ if node i stuck-at-1} \\ x \text{ if node i is fault-free} \end{cases}$$

and $\varphi: F \to R$

where, for example, $\varphi((0,x,1))$ is the faulty system:

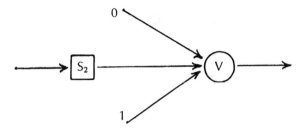

The requirement that F contain at least one benign fault is satisfied by the fault $f_0 = (x,x,x)$.

Example 4

Consider the representation scheme (S, R, ρ) of Example 2 and suppose the system in question is a PL/I program that is to compute n!, given some positive integer n. More precisely, the specification of the system is the recursive function

$$g: N \rightarrow N \qquad N = \langle 1,2,3, \ldots \rangle$$

where $g(n) = n!$

If it is assumed that any fault is possible (as long as the result is some PL/I program) and if faults are represented by their corresponding faulty programs then (S,F,φ) is a system with faults where

$$S = g$$
$$F = R$$
and $\varphi: F \rightarrow R$ is the identity function

If P is the PL/I program:

```
DECLARE (A,I,N) FIXED DECIMAL;
A = 1;
DO I = 2 TO N BY 1;
    A = A*I; End;
```

then, since $\rho(\varphi(P)) = \rho(P) = g$, P is benign when regarded as a fault and fault-free when regarded as a realization.

Life Faults vs. Birth Faults

The above examples illustrate that either life faults or birth faults (as informally distinguished in the introduction) can constitute the fault set of a system with faults. In general, the formal distinction between these two types of faults is inherent in the kind of representation scheme required.

In representing faults that occur in the process of using a system, one begins with a realization that is presumed fault-free and is concerned with faulty realizations that result from life faults. In this case, a specification *is* a realization and one chooses a representation scheme

$$(R, R, \rho)$$

where ρ is the identity function on R. Accordingly, if (S,F,φ) is a system with faults in such a scheme, then F represents a set of undesired changes in structure that can occur during the use of the system. (See Example 3) Most of the theoretical studies of fault tolerance and fault diagnosis in digital computing systems have been based on representation schemes of this type.

On the other hand, in representing faults that occur in the process of designing a system, a representation scheme (S, R, ρ) is chosen such that

(1)   The original design specification is represented by a member of the specification class S,

(2)   The possible outcomes of the design process are represented by members of the realization class R.

Since specifications differ from realizations in this case (or otherwise we would be saying that the desired outcome of the design process has already been achieved), a study of birth faults requires a representation scheme (S, R, ρ) where $S \neq R$. Accordingly, if (S,F,φ) is a system with faults in such a scheme, then F represents a set of undesired changes in structure that can occur during the design of the system. (See Example 4)

Tolerance Relations

Relative to a given representation scheme, the general concept of one system being "within tolerance" of another can also be made precise. Informally, a tolerance relation describes which realizations are "acceptable" with respect to a given specification. Formally, a tolerance relation for a representation scheme (S, R, ρ) is a relation τ between R and S such that, for all $R \in R$, $R \tau \rho(R)$. This latter condition says simply that if R realizes S then R is within tolerance of S. Thus, the most severe type of tolerance that can be imposed is the extreme case where $\tau = \rho$. A less restricted tolerance relation τ might require that $R \tau S$ whenever the behavior of realization R is the same as that of specification S. An even less severe tolerance relation might require that only certain aspects of the behavior of R need be the same as those of S.

To illustrate the concept of a tolerance relation, consider the kind of representation schemes that underlie much of the recent work on proving programs correct (see [3], for example). Here a program P is viewed as operating on input values x in some input domain X and, in those cases where the execution terminates, P yields an output value y in some output domain Y. In other words, P realizes a partial function β_p: X → Y. A program is specified by a pair of predicates φ and ψ, defined on X and X X Y respectively. φ is the "input predicate" where $\varphi(x)$ is true iff x is an input value for which an output value is desired. ψ is the "output predicate" where $\psi(x,y)$ is true iff y is the desired output value for input value x. Thus, for some given class P of (formal) programs, where P ϵ P realizes partial function β_p, the corresponding representation scheme is (S, R, ρ) where

$$S = \langle(\varphi,\psi)|\text{Dom}(\varphi) = X \text{ and Dom}(\psi) = X \times Y\rangle$$
$$R = P$$

and $\rho: R \rightarrow S, \rho(P) = (\varphi_P,\psi_P)$

   where

   $\psi_P(x)$: x ϵ Dom(β_P)
   $\psi_P(x,y)$: $\varphi_P(x)$ and y = $\beta_P(x)$

Given such a representation scheme, the following tolerance relations correspond to the concepts "correct" and "partially correct," as they are usually employed in the context of program verification.

Example 5 $(\tau_c$; "correct"):

   $P\tau_c(\varphi,\psi)$ if $\varphi(x)$ implies x ϵ Dom(β_P) and $\psi(x, \beta_P(x))$.

Example 6 $(\tau_p$; "partially correct"):

   $P\tau_p(\varphi,\psi)$ if $\varphi(x)$ and x ϵ Dom(β_P) imply $\psi(x, \beta_P(x))$.

From the above definition it is clear that "partially correct" is a weaker tolerance relation than "correct" since, in the former case, no restrictions are placed on executions which do not terminate. In particular, if no executions of P terminate (i.e., Dom(β_P) = ϕ) then P is vacuously a partially correct program.

Reliable Design of Software

The general concepts described in the previous sections suggest a number of things that should be sought in the process of investigating the reliable design of software.

(1)   Representation schemes for software systems, that is, formal descriptions of software specifications and software realizations.
(2)   Methods of describing the software design process, that is, the process of moving from a specification S to a realization R.
(3)   Descriptions of classes of birth faults that have meaningful interpretation in the context of software design.
(4)   Tolerance relations τ (for a given representation scheme) that reflect the computational needs of the user.
(5)   Software verification procedures (relative to a given tolerance relation τ), that is, procedures for establishing that R τ S.
(6)   Software diagnosis procedures, that is, procedures for establishing that R $\not\tau$ S.

In the subsections that follow, we give a brief survey of some of the research that has been performed to date in each of these areas.

Representation Schemes —1

Many of the representation schemes that have been employed to study software realiability are schemes of the type described earlier in the section on Tolerance Relations. Here the specifications are predicate pairs and the realizations are descriptions of sequential programs, usually at some relatively high level of abstraction. For example, in formalizing the assertion approach to program verification, Floyd [4] has used a scheme of this type where the realizations are "flowchart programs." Subsequent work by Manna and Pnueli [5], [6] has been based on a scheme of this type where the realizations are "recursive programs." Other more informal approaches to proving programs (see [7], for example) have considered schemes where the realizations are programs in a high-level language such as ALGOL.

Representation schemes of a somewhat more algebraic nature have also been used to study software reliability questions and, in particular, program correctness [8], [9]. Here the specifications are referred to as "state relations" and the realizations are "flow diagram programs." Again, the programs being represented are sequential as opposed to "parallel."

In the context of software reliability, much less effort has been devoted to the study of "parallel" systems composed of sequential programs which execute simultaneously and asynchronously, and which cooperate via interaction. However, beginning with Petri's formulation of an asynchronous computation model [10], a variety of description schemes have been employed to investigate parallel computation, per se, many of which could serve either as the specification class or the realization class of a representation scheme. These include parallel program schemata of the type first introduced by Karp and Miller [11] and Luconi [12], the more informal descriptions of cooperating sequential processes proposed by Dijkstra [13], and computation graphs of the type recently studied by Adams [14].

Although such descriptions might serve either as specifications or realizations, the concept of a representation scheme requires a pair of description classes S (specifications) and R (realizations) that can be related via a realization function ρ. In this regard, the process-based descriptions recently developed by Riddle [15] comprise a promising representation scheme. The specifications, in this case, are "message transfer expressions;" the realizations are "program process models;" and, given any program process model R, there is an effective procedure for determining the message-transfer-expression S such that $\rho(R) = S$.

Descriptions of the Design Process —2

If the realiability of the software design process is to be better understood, we must rely, to a certain extent at least, on knowledge of the design process itself. Early attempts to describe software design methods were motivated by a desire to obtain software systems that could be more easily analyzed and verified. This resulted in a number of contributions including the structuring techniques first described by Dijkstra [16], the iterative multi-level modeling methods of Zurcher and Randell [17], the "action clusters" of Naur [18], the stepwise refinement methods described by Wirth [19], and the top-down programming methods advocated by Mills [20].

The "structured programming" ideas contained in these various design models and methods underlie more recent work concerning the design of large software systems. For example, the idea of integrating the simulation of a system within an iterative design process [17], has been more fully developed by Graham, *et al* [21]. The iterative elaboration of a complex software system, as it proceeds through various stages of its design, has also been studied by Riddle [22], using program process models. There is a need for further development of formal descriptions of this type since they are compatible with the concept of a representation scheme and, in addition, provide precise intermediate descriptions of a system as it evolves from initial specification to final realization.

Fault Classes —3

Although attempts are being made to catalog various types of software design faults (bugs) that have been detected in the process of designing, diagnosing, or using software systems, formal descriptions of specific classes of faults are virtually nonexistent. Thus, in formally representing a software system as a system with faults (S, F, φ), faults are usually identified with the resulting faulty systems, that is, $F \subseteq R$ and $\varphi(f) = f$, for all $f \epsilon F$. Moreover, it is usually assumed that the fault class is "unrestricted" in the sense that any realization may be the result of a fault, that is, $\varphi(F) = R$. When both these assumptions are made, $F = R$, in other words, the fault class is taken to be the realization class. (See Example 4).

Even when such assumptions are not made explicitly, they are often implicit. For example, most studies of program verification make no initial assumptions regarding the nature of the program to be verified. It is assumed only that the object of verification is indeed a program according to some agreed upon definition. In other words, the result of a fault may be an arbitrary realization. If there is no interest in what causes a program to be (potentially) faulty (as is the case with most verification methods), "faults" can be identified with "results of faults" and it follows that the fault class is unrestricted.

As has been demonstrated in studies of hardware faults, much more can be done if some meaningful restrictions can be placed on the class of potential faults. A similar attempt should be made with regard to software faults, given that the design process can be appropriately formalized.

Tolerance Relations —4

As applied to software systems with birth faults, most of the tolerance relations considered to date have been developed as verification criteria for sequential programs. In particular, for representation schemes where specifications are described as predicate pairs, the tolerance relations that have been considered are τ_c ("correct") and τ_p ("partially correct"), as defined in Examples 5 and 6 (cf. [4], [5], and [6]). Tolerance relations that are essentially the same as these, although formulated somewhat differently, have also been defined for the algebraic representation schemes of Landin and Burstall [8] and Goguen [9].

For schemes that describe complex software systems, the formulation of appropriate tolerance relations appears to be much more difficult. However, the difficulty may be due primarily to the fact that the schemes themselves do not present the kind of information needed to judge tolerable behavior. This suggests that user requirements, as well as designer requirements, should be kept in mind when devising representation schemes for large software systems. It also suggests that representation schemes should be detailed enough to permit the formalization of meaningful tolerance relations.

Verification Procedures —5

In terms of the formal development of the previous sections, given a representation scheme (S, R, ρ) and given a tolerance relation between R and S, a "verification procedure" is a procedure for establishing that R τ S, for any system with faults (S, F, φ) and for any realization R ϵ $\varphi(F)$. In other words, if a verification procedure terminates with a positive outcome then the realization is within tolerance of its specification.

Verification procedures for representation schemes and tolerance relations of the type illustrated in the section entitled Tolerance Relations (Examples 5 and 6) have been studied extensively during the past eight years. An excellent review of this research is contained in a survey paper by Elspas, *et al.* [3]. This work includes the original contributions of Floyd [4] and Naur [23], informal techniques of the type advocated by London [7], the formal procedures of Manna [24], and various semi-formal (semi-automatic) techniques such as those investigated by King [25], Good [26], Elspas, *et al.* [27], and Deutsch [28].

It is not clear, however, that approaches of the type referred to above are appropriate when applied to large "parallel" software systems. Here, as we have noted earlier, the representation schemes differ considerably from those used to describe sequential programs, and tolerance relations appear to be more difficult to define. Much more work needs to be done in this regard, where the initial emphasis should focus on obtaining representation schemes and tolerance relations that will admit to feasible verifications procedures.

Diagnosis Procedures —6

As distinguished from verification, the intent of a diagnosis (testing) procedure is to detect faults which are not tolerated. More formally, given a representation scheme (S, R, ρ) and given a tolerance relation τ, a diagnosis procedure is a procedure for establishing that $R \not\tau S$, for any system with faults (S,F,φ) and for any realization $R \epsilon \varphi(F)$. In other words, if a diagnosis procedure terminates with a positive outcome then the realization is not within tolerance of its specification. Note that, in general, the lack of a positive outcome does not imply a positive verification outcome. Indeed, the latter would follow only in the case where the diagnosis procedure exhausts the class of potential faults. This has led Dijkstra [29] to remark that, in the case of software, " . . . testing can be used to show the presence of bugs, but never to show their absence."

The extent to which a system is diagnosed by a given procedure (e.g., the procedure can locate faults as well as detect them) depends on the nature of the tolerance relation τ. In the diagnosis of sequential programs, most of the procedures considered to date (see [30], for example) are based on tolerance relations that distinguish differences in input-output behavior, as opposed to differences in internal structure. However, knowledge of a program's internal structure is often used in the selection of appropriate tests. For example, consider a representation scheme where specifications are flow charts and realizations are programs written in some high-level programming language. In this context, a popular diagnosis procedure (see [31], for example) is to apply a sequence of inputs that exercise all branches of each decision element in the flowchart and, for each input, check to see whether the program's output is correct (i.e., is the same as that predicted by the flowchart). Formalizing the tolerance relation for this procedure, it follows that R τ S if, for a set X of inputs that exercise all branches of each decision element of S, the behavior of R restricted to X is the same as the behavior of S restricted to X. Thus τ is a tolerance relation of the type mentioned above.

In large software systems such as operating systems, diagnosis criteria are more elusive and diagnosis methods are more complex. Here, efforts are being made to decompose the diagnostic task by conducting tests during various stages of the system's implementation. In particular, depending on whether the system is implemented "top-down" or "bottom-up," a corresponding top-down or bottom-up diagnosis strategy can be employed. These and other diagnosis procedures are discussed in a comprehensive report by Goodenough and Eanes [32].

Conclusion

Employing the formal concept of a "system with faults," we have attempted to distinguish and clarify a number of fundamental research areas related to the reliable design of software. With respect to each area, we have reviewed some of the research already accomplished and, in certain cases, suggested topics that deserve further exploration. The areas so discussed are by no means inclusive of everything that is or will be of interest to those concerned about software design reliability. Indeed, we have tried to restrict our attention to basic questions which we feel must be answered if more interesting questions such as "How does one improve the verifiability or diagnosability of a software system?" are to be addressed effectively.

REFERENCES

1   J. von Neumann, "Probabilistic Logics," California Institute of Technology, 1952. Also published in *Automata Studies* (Edited by C. E. Shannon and J. McCarthy), Princeton University Press, Princeton, New Jersey, 1956.

2   J. F. Meyer, "A General Model for the Study of Fault Tolerance and Diagnosis," *Proceedings of the 6th Hawaii International Conference on System Sciences,* January, 1973.

3   B. Elspas, *et al.,* "An Assessment of Techniques for Proving Program Correctness," *Computing Surveys,* Vol. 4, June, 1972.

4   R. W. Floyd, "Assignment of Meanings to Programs," *Math. Aspects of Computer Science,* AMS, Providence, Rhode Island, 1967.

5   Z. Manna and A. Pnueli, "Formalization of Properties of Recursively Defined Functions," *Proc. ACM Symp. on Theory of Computation,* ACM, New York, 1969.

6   Z. Manna and A. Pnueli, "Formalization of Properties of Functional Programs," *JACM,* Vol. 17, July, 1970.

7   R. L. London, "Proving Programs Correct: Some Techniques and Examples," *B.I.T.,* Vol. 10, 1970.

8   P. J. Landin and R. M. Burstall, "Programs and Their Proofs: An Algebraic Approach," *Machine Intelligence*, American Elsevier, New York, 1969.

9   J. Goguen, "On Homomorphisms, Correctness, Termination, Unfoldments, and Equivalence of Flow Diagram Programs," *UCLA Tech. Report*, Eng. 7337, 1973.

10   C. A. Petri, "Fundamentals of a Theory of Asynchronous Information Flow," *Proc. IFIP*, Munich, 1962.

11   R. M. Karp and R. E. Miller, "Parallel Program Schemata," *Record 8th Symp. SWAT*, October, 1967.

12   F. L. Luconi, "Completely Functional Asynchronous Computational Structures," *Record 8th Symp. SWAT*, October, 1967.

13   E. W. Dijkstra, "Cooperating Sequential Processes," *Programming Languages*, Academic Press, London, 1968.

14   D. A. Adams, "A Computation Model with Data Flow Sequencing," *Tech. Report CS-117*, Computer Science Department, Stanford, 1968.

15   W. E. Riddle, "Modeling and Analysis of Supervisory Systems," Ph.D. thesis, Computer Science Department, Stanford, 1972.

16   E. W. Dijkstra, "The Structure of the T.H.E. Multiprogramming System," *CACM*, Vol. 11, May, 1968.

17   F. W. Zurcher and B. Randell, "Iterative Multi-Level Modeling—A Methodology for Computer System Design," *IFIP 1968*, Edinburgh, 1968.

18   P. Naur, "Programming By Action Clusters," *B.I.T.*, Vol. 9, 1969.

19   N. Wirth, "Program Development by Stepwise Refinement,"*CACM*, Vol. 18, April, 1971.

20   H. Mills, "Top-Down Programming in Large Systems," *Debugging Techniques in Large Systems*, P-H, New Jersey, 1971.

21   R. M. Graham, *et al.*, "A Software Design and Evaluation System," *CACM*, Vol. 16, February, 1973.

22   W. E. Riddle, "A Design Methodology for Complex Software Systems," *Proc. 2nd Texas Conf. on Computing Systems*, Austin, 1973.

23   P. Naur, "Proof of Algorithms by General Snapshots," *B.I.T.*, Vol. 6, 1966.

24   Z. Manna,"Properties of Programs and the First-Order Predicate Calculus," *JACM*, Vol. 16, April, 1969.

25   J. C. King, "A Program Verifier," Ph.D. thesis, Carnegie-Mellon University, 1969.

26   D. I. Good, "Toward a Man-Machine System for Proving Program Correctness," Ph.D. thesis, University of Texas, 1970.

27   B. Elspas, *et al.*, "Research in Interactive Program-Proving Techniques," SRI Project 8398, Menlo Park, California, 1972.

28   L. Deutsch, "An Interactive Program Verifier," Ph.D. thesis, University of California, Berkeley, 1973.

29   E. W. Dijkstra, "Structured Programming," *Software Engineering Techniques*, NATO, Brussels, 1970.

30   W. C. Hetzel (Ed.), *Program Test Methods*, P-H, New Jersey, 1973.

31   M. R. Paige, "Software Testing: An Overview," *Digest 4th Symp. on FTC*, Urbana, Illinois, 1974.

32   J. B. Goodenough and R. S. Eanes, "Program Testing and Diagnosis," Tech. Report, SofTech, Inc., Waltham, Mass., 1973.

A FUNCTIONAL APPROACH TO FAULT ANALYSIS OF LINEAR SYSTEMS

M. N. Ransom

R. Saeks

Introduction

With the rapid advance in the complexity and size of modern electronic systems has come the need for automated aids for locating component failures in these systems. Former techniques which were basically heuristic in nature and relied heavily on the experience and "intuition" of the maintenance personnel can no longer be relied upon. This is especially true in situations where the system down time must be minimized as is the case of telephone communication systems. For the case of digital systems a number of automatic fault isolation techniques have been developed and are integral parts of most large-scale digital systems. On the other hand only small advances have been achieved so far in developing a theoretical basis for linear fault analysis. It is the belief of the authors that much of this difficulty has been brought about because the input-output-state and node-loop equation formulation upon which past attempts have been based fail to adequately show the relation which exists between the overall response of a system and the individual component responses.

This paper explores a unique approach to fault analysis based on a system modeling technique wherein the observable system behavior is expressed explicitly as a function of the internal component responses. It has been shown that this function is determined entirely by the connectivity structure of the system and for that reason has been referred to as the "connection function" of the system. This function has received considerable attention in a number of recent papers both as a theoretical tool and in such

M. N. Ransom is with Bell Laboratories, Naperville, Illinois.

R. Saeks is an Associate Professor in the Departments of Electrical Engineering and Mathematics, Texas Tech University, Lubbock, Texas.

This research supported in part by ONR Contract 73-A-0434-0003 and NSF Grant GK-31808.

wide areas as sensitivity analysis and computer aided design [1,2,3,]. In the present discussion it will be shown that the problem of fault analysis of linear systems can be expressed in terms of the inverse of the connection function. It is the author's belief that this approach holds much potential in establishing some of the theoretical foundation required for developing effective linear fault analysis techniques.

The Connection Function as a Fault Analysis Tool

Before presenting this functional approach to fault analysis it is first useful to obtain a definition of fault analysis and to contrast fault analysis with so-called fault detection and fault isolation techniques. The term fault detection usually refers to techniques which determine whether a system or subsystem being examined is working satisfactorily in all respects. Fault isolation, on the other hand, refers to techniques used to determine which component in a system is faulty—although not necessarily what is wrong with the component, only that its response differs from the desired response. Fault analysis, which will be considered here, seeks not only to determine which components have responses differing from their desired responses but also tries to establish what their new responses are. Although such a detailed analysis is usually more than is needed in most repair decisions, fault analysis will be examined here because related theory is more easily developed when dealing with linear systems and because, in most cases, the requirements on a linear system that allow fault isolation are the same as those that allow fault analysis.

For our purposes we will define the problem of fault analysis as that of determining from measurements made at available terminals or test points of a system the input-output response of each of the system's components. Note then that the only way we are allowed to make measurements on the system is to apply signals to some of these terminals and measure the resulting terminal response. Recall that the system matrix for a linear system is defined as that matrix function of frequency that when applied to a set of inputs yields the corresponding system response. Thus it is easily seen that all information obtainable through measurements could have been derived from the system matrix of the failed system (were it known) defined over all available inputs and outputs (including those provided for testing purposes only). Furthermore it is easily shown that the system matrix at any frequency can be evaluated by applying appropriate inputs and measuring the corresponding outputs. Thus we say that the totality of the system behavior that is measurable from the available terminals of a system is the system matrix.

From the above discussion we have that the study of the relationship which exists between the measurable system behavior and the component responses becomes that of finding the relationship between the system matrix and the component responses. In the next section we will show that this relationship can be expressed as

$$S = f(Z) \tag{1}$$

where S is the system matrix relating inputs and outputs of the failed system, Z is a block diagonal matrix whose diagonal elements are matrices of operators modeling each of the system's components and f, termed the "connection function" of the system, is a function, independent of frequency, determined entirely by the connectivity structure of the system.

From equation (1) we have the problem of a fault analysis (where all measurements are made at a single frequency) that has been reduced to finding the inverse of the connection function in which case we have

$$Z_i = f^{-1}(S_i) \tag{2}$$

where S_i is the system matrix measured at the frequency ω_i and Z_i is the components' response exhibited at that frequency. Thus from the above formulation we can immediately assert the following:
Proposition 1

For the case where measurements are made at a single frequency, fault analysis of a linear system is possible for any exhibited set of component responses if and only if the connection function possesses an inverse.

Here the connection function is defined over all system terminals at which measurements are made. Strictly speaking we require only that the connection function possess a left inverse, that is for any Z_1 and Z_2, $f(Z_1) = f(Z_2)$ if and only if $Z_1 = Z_2$. However, since we are only considering measured system responses, we are guaranteed that f is onto, i.e., there exists a set of component responses that would realize the measured system response (scilicet, the actual component responses). Thus we can impose the more general requirement that f be invertible.

In general the above requirement cannot be relaxed for the case where measurements are made at multiple frequencies unless the type of responses allowed by the components are restricted in such a way that information obtained at one frequency can be applied to responses at other frequencies. This can be done by assuming a failure model for each of the components. These failure models should be sufficiently general so as to include any possible failure mechanism, yet not be so general as to require an inordinate number of measurements to perform the fault analysis. Use of failure models serves to restrict the deviation from normality that a component can be expected to make, yet also allows us to include information on the types of failures normally expected. For example, a failed resistor will still behave as a positive resistor and capacitors usually fail by developing either a series resistance or parallel conductance, passive elements remain passive after failure, and so forth. Determining effective failure models for fault analysis is itself an important area for further study and will not be discussed here.

By substituting a failure model for each component of the system we have that the actual response exhibited by the components, Z, can be written as

$$Z = g_\omega(C) \tag{3}$$

where C is a matrix of unknown parameters in the failure models and g_ω is a frequency dependent function of C. Thus for the multifrequency case the problem of fault analysis is to find a unique set of parameters C' such that the expression

$$S = f[g_\omega(C')] \tag{4}$$

is satisfied for each frequency at which S is measured. We can thus state the following:

Proposition 2

For systems whose response can be measured at any number of frequencies, fault analysis is possible for any exhibited set of component responses if and only if the equation

$$f[g_\omega(C_1)] = f[g_\omega(C_2)] \tag{5}$$

is satisfied for all ω only if $C_1 = C_2$.

If, for the single frequency case, proposition 1 is not satisfied or, in the multifrequency case, proposition 2 is not satisfied then there exists more than one plausible set of component responses which would produce the observed system behavior. In these cases further test points would be required to establish the actual component responses. In situations where no other test points are available we may wish to make the "best guess" of the component responses consistent with the measurements that are available. A portion of this problem was considered by the authors in a previous paper [4]. There the most probable set of component responses was considered to be, of all sets of responses which would produce the observed system response, that set closest to the nominal responses. This would correspond to failures caused by components drifting out of tolerance due to age, thermal drift, or component tolerances. Thus in terms of the connection function the problem of "approximate" fault analysis becomes that of finding all sets of component values C' satisfying the relation

$$S_1 = f[g_1(C')]$$

$$\cdot \ \cdot \ \cdot \ \cdot \ \cdot \ \cdot \tag{6}$$

$$S_k = f[g_k(C')]$$

that set C^* for which $\|C_0 - C'\|$ is minimized. Here measurements are made at k different frequencies $\omega_1 \cdots \omega_k$, $g_i = g_{\omega_i}$ and C_0 is the nominal value of the failure model parameters. An explicit matrix expression which closely approximates the above nonlinear optimization problem for the single frequency case was derived in the aforementioned paper and will be described in a later section.

Other measures of what constitutes the best guess for the approximate fault analysis problem can be applied and may be considered in future studies. For example, to handle the case of catastrophic failures an appropriate optimization might be to determine that set of component responses which would exhibit the observed system response with the least number of non-nominal response components.

Although an abstract formulation of the fault analysis problem in terms of the connection function as described above is conceptionally pleasing, it is of no practical value unless one has a "hard" and computationally viable representation of the connection function. Fortunately such a representation exists and will be presented in the following section.

The Connection Function

The idea of expressing the response of a system as a function of its components is not new. Indeed this is done whenever one writes the transfer function of a circuit explicitly in terms of its internal component parameters, and such an approach is the basis of the various symbolic network analysis programs. Unfortunately the *ad hoc* nature of these formulations has in the past precluded their full exploitation. A more unified approach to expressing this relationship and one whose mathematical structure is sufficiently general and succinct to lend itself to functional analysis was recently proposed by the authors [1,2]. This unified approach is based on the component-connection model adopted in a number of recent studies to describe large scale dynamical systems [6-9]. The principle advantage the component-connection model has in large scale system theory is that it allows the constraints placed on a system by each of its subsystems to be separated from those placed on the system by the interconnections between the subsystems. This model though is sufficiently general as to be applied to systems of any size and whose subsystems can simply be circuit elements.

The development of the component-connection model is illustrated in Figures 1(a) and 1(b) [page 133]. Here u and y are vectors containing the inputs and outputs, respectively, of the overall system, and a and b are vectors whose elements are the inputs and outputs, respectively, of each of the individual components of the system. The hatched area then represents the system components and the shaded area represents the interconnections between the components and between the system inputs and outputs and the components. Here the overall system is described by a transformation S

which maps any permissible system input u into the resulting output y *via*
$$y = S \cdot u \tag{7}$$
and the components are described by the transformation Z which maps the component input variables a into the component output variables b *via*
$$b = Z \cdot a \tag{8}$$
If we now redraw Figure 1(a) as shown in Figure 1(b) the components and the connections appear separated and the system can be seen to be decomposed into two boxes: the components described by equation (8) and the "donut shaped" connection box whose inputs are u and b and whose outputs are y and a. In that the interconnections of a system are characterized by linear algebraic equations (Kirchoff laws, scalar and adder equations, etc.) it is reasonable to model the connection box by the matrix equation

$$\begin{bmatrix} a \\ y \end{bmatrix} = \begin{bmatrix} L_{11} & L_{12} \\ L_{21} & L_{22} \end{bmatrix} \begin{bmatrix} b \\ u \end{bmatrix} \tag{9}$$

Equations (8) and (9) have been termed the component-connection system model. Conditions for the existence for the connection equations and algorithms for generating the component-connection equations for an arbitrary system have been studied and computer implementations are available [9-12].

Solving equations (7), (8) and (9) simultaneously we obtain the desired connection function relating the input-output behavior of a system to its components' response, *viz.*
$$S = f(Z) = L_{22} + L_{21}(1 - ZL_{11})^{-1} ZL_{12} \tag{10}$$
Note that even for linear systems the relationship between component responses and the overall system response is, in general, nonlinear. Note further that the connection function can be decomposed into three functions, a nonlinear, invertible function
$$R = g(Z) = (1 - ZL_{11})^{-1} Z \tag{11}$$

a linear, in general noninvertible function

$$Q = h(R) = L_{21} R L_{12} \tag{12}$$

and a shifting function

$$S = k(Q) = L_{22} + Q \tag{13}$$

Thus the only noninvertibility of the connection function occurs in the linear portion of the decomposition. This fact will be shown to be quite useful in a later section.

Conditions For Linear Fault Analysis

First consider the case where measurements are only made at a single frequency. By proposition 1 fault analysis can be achieved only if the connection function possesses an inverse. In the case where all elements of Z must be determined, applying the invertibility condition to equation (10) yields [13]:

Proposition 3
Single frequency fault analysis of a linear system for which all elements of the component matrix are unknown can be achieved if and only if both L_{21}^{-L} and L_{12}^{-R} exist, in which case

$$Z_i = L_{21}^{-L}(S_i - L_{22})L_{12}^{-R} [1 - L_{11}L_{21}^{-L}(S_i - L_{22})L_{12}^{-R}]^{-1} \qquad (14)$$

where -L and -R denotes the left and right inverses, respectively, S_i is the system response measured at frequency ω_i and Z_i is the components' response at that frequency.

In the more usual situation most of the off-diagonal elements of Z are known to be zero (even during system failure). It is easily shown that through matrix manipulations a new set of connection equations can be written for which all the nonzero elements of Z are placed along the diagonal. Therefore we can assume without loss of generality that Z is diagonal. In a previous paper the following properties were shown to exist [3]:

Proposition 4
Single frequency fault analysis of a linear system where the sought component matrix is diagonal can be achieved if any of the following conditions are satisfied:
(1) L_{11} is diagonal and $(L_{21}^t \odot L_{12})^{-L}$ exists.
(2) L_{21}^{-L} exists and $L_{21}^{-L}(S_i - L_{22})$ has no zero rows.
(3) L_{12}^{-R} exists and $(S_i - L_{22})L_{12}^{-R}$ has no zero columns.
Here t indicates transpose and $\odot$ denotes the Kronecker dot product [16]. The proofs for the above properties are constructive. That is, given that the condition is satisfied, a technique for establishing component responses from external measurements can be demonstrated.

An interesting technique for linear fault analysis which can be applied in the general case where none of the above conditions are necessarily satisfied and for the multifrequency case was recently proposed in a previous paper [14]. In that paper it was shown that through suitable manipulation equation (10) can be placed in the form

$$\psi_i \xi_i = \gamma_i \qquad (15)$$

where ψ_i and γ_i are a matrix and vector, respectively, whose elements are determined by the connection matrices and measured system response at frequency ω_i. Here ξ_i is a vector whose elements are products of the sought component responses taken one at a time through n at a time, where n is the number of unknown component responses. Since the first n elements of ξ are our sought responses, fault analysis consists of reducing the above equation to the form

$$\begin{bmatrix} I & 0 \\ X & Y \end{bmatrix} \xi_i = \gamma'_i \qquad (16)$$

where I is an n x n identity matrix and the unknown component values are thus the first n elements of γ'_i. In cases where only a few rows of the upper partition can be achieved, the determined responses corresponding to these rows can be substituted into ξ_i and the row reduction continued. This process of row reduction and substitution is continued until the form of equation is achieved or until a row reduction yields no additional responses.

For the case where measurements are made at a number of different frequencies, we assume that each component response (or each element of the failure models for the components) can be written as

$$z = b(\omega)c \qquad (17)$$

where c is the unknown component (or parameter) value and $b(\omega)$ is a known function of frequency. Substituting (17) into ξ we have

$$\xi_i = B_i \zeta \qquad (18)$$

where B_i is a known matrix function of frequency evaluated at ω_i, and ζ is the same vector of products as ξ except with the component element values substituted for component responses. We can now combine the observations made at the different frequencies into the one equation

$$\begin{bmatrix} \psi_1 & B_1 \\ \psi_2 & B_2 \\ \vdots \\ \psi_k & B_k \end{bmatrix} = \begin{bmatrix} \gamma_1 \\ \gamma_1 \\ \\ \gamma_k \end{bmatrix} \qquad (19)$$

The above equation is then reduced to the form of equation (16) except now the additional number of rows increases the chance that the n rows of the upper partition can be achieved.

The above technique has the interesting property that it allows some of the component values to be determined where only enough measurements are available to determine these but not other component values. Also in situations where some of the component values are known or suspected to be correct because of much higher reliability, these values can be substituted during the reduction process.

Approximate Fault Analysis

We now briefly consider the situation where insufficient measurements are available to determine exactly the internal component responses. Here we wish to determine the most probable component responses consistent with those measurements that are available. This, for instance, could tell us which components should be tested first. As pointed out earlier we will consider for the present discussion the most probable solution to be the "closest to nominal" solution which yields the observed system behavior. This problem for the single frequency case was considered in a previous paper and the solution technique employed is summarized below.

Let Z_0 be the nominal component responses at the frequency at which measurements are taken and S_m be the measured system response. The problem is then to find, of all matrices Z' satisfying

$$S_m = L_{22} + L_{21}(1 - Z'L_{11})^{-1}Z'L_{12} \qquad (20)$$

that matrix Z^* which minimizes $\|Z' - Z_0\|$. A solution which satisfies equation (20) and which closely approximates Z^* is found by noting that the noninvertible portion of the connection function is linear (equation 12). Thus any solution $Z = g^{-1}(R)$ (equation 11) which satisfies

$$S_m = L_{22} + L_{21}RL_{12} \qquad (21)$$

also satisfies equation (20). Let $R_0 = g(Z_0)$. A solution to (21) which minimizes $\|R - R_0\|$ is easily obtained through linear optimization and is given by

$$vec(R') = vec(R_0) + (L_{12}^t \otimes L_{21})^{-G} vec(S_m - S_0) \qquad (22)$$

where S_0 is the nominal system response $f(Z_0)$, $\otimes$ denotes the kronecker cross product [15], -G denotes the generalized or pseudo-matrix inverse and vec is an operator which forms a vector by placing the columns of the matrix one successively under the other. There is no guarantee, however, that minimizing $\|R - R_0\|$ also minimizes $\|Z - Z_0\|$. On the other hand it can be made to nearly do so by introducing a suitable linear invertible weighting function p into the optimization on R such that we are minimizing $\|p(R) - p(R_0)\|$. It is shown in the aforementioned paper that the correct weighting function is given by the derivative of $h(Z)$ evaluated at Z_0. Fortunately $h(Z)$ is of such a form that an explicit matrix expression for its inverse exists and a closed form matrix expression for the sought optimization is given in reference 4. This technique has been found to yield surprisingly close estimations for component responses with only a few external measurements available.

No similar solution for multifrequency approximate fault analysis has been determined. On the other hand, because the special nature of the connection function allows explicit matrix expressions for its derivative, nonlinear optimization can be effectively applied to the problem. Details of such an approach are given in reference 3.

Conclusion

The suggested technique of approaching linear fault analysis by allowing the interconnections of a system to be seen as a function mapping component responses into the system response has been found to be useful both theoretically and as a means of developing viable fault analysis procedures. Studies are now underway to extend this approach to both the nonlinear case and to digital fault analysis.

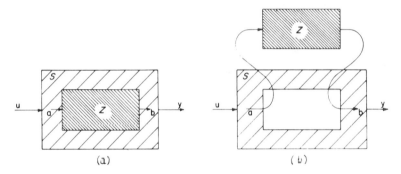

Figure 1. Development of the component connection model

REFERENCES

1. M. N. Ransom and R. Saeks, "The Connection Function - Theory and Application" *Inter. Journal of Circuit Theory and Application,* Vol. 3, 5-21 (1975).

2. M. N. Ransom and R. Saeks, "A Functional Approach to Large-Scale Systems", *Proc. 10th Allerton Conf. Circuit and System Theory,* Univ. of Illinois, 48-55 (1972).

3. M. N. Ransom, "A Functional Approach to the Connections of a Large-Scale Dynamical System," Ph.D. Thesis, Univ. of Notre Dame, 1973.

4. M. N. Ransom and R. Saeks, "Fault Isolation with Insufficient Measurements, *IEEE Trans. Circuit Theory,* CT-20, 416-417 (1973).

5. S. P. Singh and R. W. Liu, "Existence of State Equation Representation of Linear Large - Scale Dynamical Systems," *IEEE Trans. Circuit Theory*, CT-20, 239-246 (1973).

6. F. Liu, "Stability of Large - Scale Dynamical Systems," Ph. D. Thesis, National Taiwan Univ., 1971.

7. M. H. Richardson, R. J. Leake and R. Saeks, "A Component Connection Formulation for Large - Scale Discrete - Time System Optimization," *Proc. 3rd Asilomar Conf. Circuits and Systems*, 665-670 (1969).

8. R. Saeks, "Fault Isolation, Component Decoupling and the Connection Groupoid," Research Reports 1970 NSAS - ASEE Summer Faculty Fellowship Program, Auburn Univ., 505-534 (1970).

9. S. P. Singh, "Structural Properties of Large-Scale Dynamical Systems," Ph.D. Thesis, Univ. of Notre Dame, 1972.

10. N. Prasad and J. Reiss, "The Digital Simulation of Interconnected Systems," *Proc. Conf. Inter. Assoc. of Cybernetics*, Belgium, 1970.

11. H. Trauboth and N. Prasad, "MARSYAS - A Software System for the Simulation of Digital Systems," *Proc. Spring Joint Comp. Conf.*, 223-235 (1970).

12. S. P. Singh and H. Trauboth, "MARSYAS," IEEE Circuits and Systems Soc. Newsletter, 7 (1973).

13. R. Saeks, S. P. Singh and R. W., Liu, "Fault Isolation via Components Simulation," *IEEE Trans. Circuit Theory*, CT-19, 634-640 (1972).

14. M. N. Ransom and R. Saeks, "Fault Isolation via Term Expansion," *Proc. 3rd Pittsburgh Symp. on Modeling and Simulation*, Univ. of Pittsburgh, Vol. 4, 224-228 (1973).

15. H. Neudecker, "Some Theorems on Matrix Differentiation with Special Reference to Kronecker Matrix Products," *Amer. Stat. Assoc. Journal*, 953-963 (1969).

16. C. G. Khatri and C. R. Rao, "Solutions to Some Functional Equations and Their Applications to Characterization of Probability Distribution," *Sankha*, Ser. A, Vol. 30, 167-180 (1968).

17. R. Penrose, "On the Best Approximate Solutions of Linear Matrix Equations," *Proc. Cambridge Philos. Soc.*, Vol. 31, 406-413 (1956).

FAULT PREDICTION — TOWARD A MATHEMATICAL THEORY

S. R. Liberty, L. Tung, and R. Saeks

Introduction

The fault analysis process for a digital device is usually divided into two steps: fault detection and fault location. In the former one decides whether or not the device is operating correctly, and in the latter one determines which component(s) is the cause of faulty operation. Both of these decisions are unambiguous. A digital device is either working or not, its components are either working or not; there is no ambiguity.

Unlike the digital case, in the fault analysis of an analog device one cannot make a black and white decision. There are many shades of gray between the black and white of nominal behavior and outright failure. Although this gray area greatly complicates the decision process it also makes possible a third aspect of the fault analysis process for analog devices—fault prediction. That is, one tests a device at periodic maintenance intervals. If a component is out of tolerance it is replaced immediately; whereas if its value is nearly nominal the component may be assumed to be good. On the other hand for those components in the gray region near their tolerance limit one can attempt to predict (on the basis of estimated component values at the present and prior maintenance intervals) whether or not the component will go out of tolerance before the next scheduled maintenance. The component can then be replaced before it actually fails.

Of course, there is little hope of predicting component failure due to random external effects (improper operation, lightning, etc.), but there is significant experimental evidence to suggest that when electronic components fail due to permanent overstress (high temperature, continued overload operation, material fatigue, etc.) that their parameters change sufficiently slowly so as to predict the time at which they will go out of tolerance by statistical trending techniques [1,2,3].

The purpose of the present paper is to describe the rudiments of a fault prediction algorithm for use in conjunction with a periodic maintenance system. The algorithm is divided into three steps. First, one must develop

S. R. Liberty, L. Tung, and R. Saeks are with the Department of Electrical Engineering, Texas Tech University, Lubbock, Texas 79409.

This research supported in part by the Office of Naval Research Contract ONR-73A0434-0003.

techniques for estimating the internal component values of an analog device from measurements of the device at externally accessible test points at each periodic maintenance. Second, the resultant sequence of estimated component values, possibly combined with a priori reliability data, is used to estimate the expected time of failure. Finally, a decision is made as to whether or not to replace the component. The decision is based on the component's probability of failure before the next scheduled maintenance, the cost of replacement, and the cost of an on-line failure.

The problem of estimating device component values from external data has been extensively studied by one of the authors [4-8] and will not be considered here. See reference 6 for a review of this work. The present paper will therefore be devoted to the latter two steps in the fault prediction process. After a brief description of the periodic maintenance philosophy a technique for estimating the failure time for a component from a sequence of its estimated component values will be described. This is a statistical technique based on a least squares fit and is predicated on the assumption that the component parameters change sufficiently slowly to allow approximation by a low order polynomial. Finally, we will consider the problem of making the replacement decision given an estimated failure time and its higher order statistics, a priori reliability data, and the various costs involved. For this purpose actuarial techniques will be invoked [9].

A Periodic Maintenance System

By a periodic maintenance system for an electronic device we mean an algorithmic procedure, probably automated, for testing the device at fixed intervals (monthly, quarterly, annually, etc.) to determine whether it is operating properly at that time, to determine the source of the difficulty if it is not operating properly, and to predict any faults which may be expected to occur before the next scheduled maintenance. That is, a periodic maintenance system should be capable of fault detection, fault isolation, and fault prediction.

To implement a periodic maintenance system for an electronic device we assume that the device is equipped with a test jack from which sufficiently many measurements may be made to estimate accurately all relevant internal component parameters. Procedures are discussed in references 4 through 8 for choosing the test measurements and designing the software for processing the resultant data to obtain the desired component value estimates. These procedures, although still incomplete, are in a sufficiently well developed state to be implemented in a periodic maintenance system, and as such, they will not be considered in detail here. Once obtained, these component value estimates suffice to achieve both fault detection and fault location. In particular, if one assumes that the components within the electronic device are correctly interconnected*, the overall device will be operating properly if all components are within tolerance. If the device is faulty, knowledge of which components are out of tolerance will serve to locate the fault.

To carry out the fault prediction task we assume that the device has been tested at a sequence of periodic maintenance intervals (say at times t_0, $t_1, \ldots, t_n$), and that estimates for each component value have been obtained at each such maintenance period. As such, at the nth maintenance period we have a sequence of values $v_0, v_1, \ldots, v_n$ for each component which may be extrapolated by appropriate statistical techniques to predict future component failures. In addition a priori component lifetime data may be employed in the prediction process. This data is then used to estimate the probability of component failure before the next scheduled maintenance and is combined with appropriate cost figures to make a replacement decision.

The above described maintenance system, which is predicated on the determination of each component parameter at each maintenance interval whether good or bad, is capable of simultaneously achieving fault detection, fault location, and fault prediction. This is unlike many classical maintenance systems wherein one simply makes a binary decision as to whether various components are good or bad at the time at which they are tested. Of course, obtaining an accurate estimate of each component value at each maintenance interval will demand increased device measurements and/or software over that required for a simple binary decision.

Estimating Failure Time

Our estimate for the failure time of a component will be based on two sets of data: manufacturers, a priori lifetime statistics and a time sequence of actual component values measured at the various maintenance intervals. The final failure time estimate will be taken to be a weighted average of estimates obtained from these two data sources. Of course, the weight factors employed will vary with our confidence in the data. When a component is first installed one has little information on the actual device and must rely heavily on a priori lifetime data (which is based on the collective of all similar devices but not on the specific device under test). After one has obtained data on the actual device in its own environment at several periodic maintenance intervals, however, one can then use this information with greater confidence in estimating failure time. This will shift the weight factor from a priori to

---

*In a periodic maintenance system it is reasonable to assume that the device was in good working order when put into service in which case future faults would be due to component failure rather than improper interconnection. Of course, such an assumption is not valid for a test system intended for use at initial assembly.

measured data. The use of two such data sources for failure time is reminiscent of the "so-called" credibility premium computation in actuarial theory [9, 10], and indeed, the required weight factors are computed via actuarial techniques.

Notationally, we will assume that a device has been tested at times t_0, $t_1, \ldots, t_n$ and that the estimated values for a particular component parameter have been found to be $v_0, v_1, \ldots, v_n$ at these times. Assuming that the upper and lower tolerance limits for the component are u and w, respectively, we desire to estimate the time $g_n > 0$ at which the component will go out of tolerance and its variance μ_n. We will also denote by g_0 and μ_0 the expected failure time and its variance for the component obtained from a priori lifetime data.

Our approach to estimating g_n will be via a least squares fit of the points (t_i, v_i), $i = 0, 1, \ldots, n$ by a second order polynomial. Once such a polynomial fit has been obtained, say $a_2 t^2 + a_1 t + a_0$, we then use the binomial formula to estimate the time at which the component value will cross its tolerance limit. Although this would appear to be an extremely naive estimate, and indeed it is, there are a number of arguments in favor of such an approach.

(i) It has often been observed by maintenance engineers that components failing due to permanent stresses tend to evolve gradually from nominal to their failed state. Indeed, for electronic components these observations have been verified by several Soviet reliability engineers [1,2,3] as well as our own, still incomplete, experiments [11]. In fact, Leont'yev [1,2] suggests that a second order function of time will suffice to model most electronic component failures. This is not to imply that there is any physical basis for a second order failure characteristic but simply implies that in the relatively small range between a component's nominal value and its tolerance limit the failure curve is sufficiently smooth to be approximated accurately by a second order fit.

(ii) Given n measurements of the component value, computationally one could as readily fit an (nth) order polynomial as a second order polynomial. The use of the low order polynomial, however, will tend to suppress noise and other random effects which might be exaggerated by a higher order fit. Hence, given the expected smoothness of the curve, a low order fit should be superior to a high order polynomial fit.

(iii) Although we would like g_n to be estimated as accurately as possible given the ultimate application of making a replacement decision, we require high accuracy only when g_n-t_n is small (i.e. failure is expected in the near future). Since this is the only case when a decision is imminent, i.e., if g_n-t_n is large, failure is not expected in the near future. Therefore even if our estimate of g_n is inaccurate, this error will not effect our replacement decision. As such, our polynomial approximation need be accurate only over a small time interval which is readily achieved by a second order fit.

(iv) Finally, a second order fit offers significant computational advantages over a high order fit. The polynomial coefficients may be computed via the inversion of a 3 by 3 matrix which being independent of the data may be precomputed [12]. Similarly, the second order fit permits g_n to be computed via the binomial formula, rather than by a polynomial zero finder, which would be required for a high order approximation.

Given the second order approach the mathematics of computing g_n is straightforward. Hence, we simply sketch the required manipulations here and refer the reader to reference 11 for the details. To compute the polynomial coefficients a_0, a_1, and a_2 we let p_0, p_1, and p_2 be any three linearly independent second order polynomials and compute the inner products

$$h_i = \sum_{k=0}^{n} v_k p_i(t_k) \quad i = 0, 1, 2$$

and

$$g_{ij} = \sum (t_k)^i p_i(t_k) \quad i, j = 0, 1, 2$$

The coefficients are then given by the Gram equation [12]

$$\begin{bmatrix} a_0 \\ a_1 \\ a_2 \end{bmatrix} = \begin{bmatrix} g_{00} & g_{01} & g_{02} \\ g_{10} & g_{11} & g_{12} \\ g_{20} & g_{21} & g_{22} \end{bmatrix}^{-1} \begin{bmatrix} h_0 \\ h_1 \\ h_2 \end{bmatrix}$$

where the Gram matrix, being independent of data, need be inverted only once.

Once the approximation polynomial $a_2 t^2 + a_1 t + a_0$ has been computed, we need only solve the second order equations

$$a_2 t^2 + a_1 t + a_0 = u$$

and

$$a^2 t^2 + a_1 t + a_0 = w$$

to determine the time at which the component will go out of tolerance. If we make the reasonable assumptions [3] that a_0 lies within the tolerance range (if it doesn't the component should have been replaced long ago) and that the failure curve is monotonic for positive values of time (i.e., a_1 and a_2 are the same sign), then the process is further simplified. In particular, if a_1 and a_2 are positive the failure curve is increasing and we may use the formula

$$g_n = \frac{-a_1 + \sqrt{a_1^2 - 4a_2(a_0 - u)}}{2a_2}$$

whereas if a_1 and a_2 are negative the failure curve is decreasing and we may use the formula

$$g_n = \frac{-a_1 - \sqrt{a_1^2 - 4a_2(a_0 - w)}}{2a_2}$$

to give us the positive value of time at which the failure curve will cross the tolerance limit. In both cases the discriminant is assured to be positive via our assumed inequalities and g_n must be greater than t_n if the component is still within tolerance at time t_n (if not it would be replaced at the nth maintenance period and no fault prediction would be needed).

The detailed statistics for the estimator g_n are developed in reference 11. Here we simply note that, within a first order approximation, the estimator is unbiased and its variance decreases linearly with n.

$$\mu_n = 0(1/n)$$

These computations are based on the assumption that the actual failure curve is second order and that our measurements differ from the actual curve by an additive white noise term.

Finally, we need to consider the problem of combining the measured estimate g_n with the a priori lifetime g_0. For this purpose we will take our combined estimate to be the weighted sum

$$g = \alpha g_0 + (1 - \alpha)g_n \quad 0 \leqslant \alpha \leqslant 1$$

Now if we assume that g_0 and g_n are independent (since they are obtained from totally unrelated sources of data), then it is readily shown [9,11] that

$$\alpha = \frac{\mu_n}{\mu_0 + \mu_n}$$

We note that μ_0 is independent of n whereas $\mu_n = 0(1/n)$. Hence as the number of periodic maintenance intervals increases our estimator places more and more confidence on measured data, completely disregarding a priori lifetime data in the limit.

The Replacement Decision

Once our combined failure estimate, g, and its variance, μ, (and possibly other statistics) have been computed it remains to make a replacement decision. This should be based on the probability of the component failing between time t_n and t_{n+1} (the next scheduled maintenance), the cost of replacing the component, and the cost of an on-line failure. One possible replacement criterion which may be readily implemented via standard actuarial methods [9] is based on the relative costs of insuring the original component against failure until the next scheduled maintenance, versus the cost of insuring a replacement component.

That is, we compute the premiums P_0 and P_n for insuring the old and new components against failure during the interval $[t_n, t_{n+1}]$. Both premiums reflect the probability of component failure during the maintenance interval and the cost of such a failure. For the old component we have measured data as well as a priori lifetime data and may compute P_0 as a function of the four parameters t_n, t_{n+1}, g, and μ. On the other hand for the new component there is no measured data and our premium computation will be based on the a priori parameters t_n, t_{n+1}, g_0, and μ_0.

Finally, the cost of replacement should be considered. For this purpose we do not desire to charge the entire cost of the replacement component to the maintenance system but only that part of the component's potential lifetime which is wasted because of the "conservatism" of the replacement criterion. The portion of a component's lifetime which is actually used should be charged to the operating cost of the device. As such, let C denote the total cost of a component throughout its expected lifetime, g_0, (including initial cost, depreciation, interest, installation, etc.). Now let

$$C_r = \frac{g_{11} - t_{11}}{g_0} C$$

represent that portion of the component cost which would be wasted if the component were replaced at the nth maintenance interval rather than waiting for it to fail (at g_n).

Given these three parameters we may make our replacement decision by comparing the cost of insuring the old component until the next scheduled maintenance, P_0, with the cost of replacing the component plus insurance on the new component, P_n. We then compute

$$P_0(t_n, t_{n+1}, g_n, \mu_n)$$

and

$$P_n(t_n, t_{n+1}, g_0, \mu_0) + C_r(t_n, g_n, g_0, C)$$

and replace the component whenever the latter is less than the former. This criterion is, given the availability data, optimal over the time interval $[t_n,$

$t_{n+1}]$, though it is not known whether or not it is globally optimal (i.e., minimizes the total maintenance cost over the lifetime of the device).

Conclusions

Our purpose in the preceding has been to describe the rudiments of a periodic maintenance system capable of fault detection, fault location, and fault prediction. The fault prediction scheme is based wholly on the presumed smoothness of the failure curve and is therefore independent of properties of any specific component.

A maintenance system based on these ideas is presently being simulated to determine the viability of the scheme. For this purpose components with smooth (plus noise) but non-quadratic failure curves are being used to show that failure prediction can be achieved without a priori knowledge of the failure curve.

Finally, several variations on the above scheme are being investigated. In particular the use of weighted least squares and spline fits are being considered to cope with components whose failure trends change while in service.

REFERENCES

1. Leont'yev, L. P., *Introduction to the Theory of Reliability of Radio-Electronic Apparatus*, Riga, AN Latv., 1963.

2. Leont'yev, L. P., and A. M. Margulia, "Reliability and Length of Service of Certain Vacuum Tubes," Proc. of the Inst. of Electronic and Computational Technology of the Academy of Sciences of the Latvian SSR, Vol. 5, 1963.

3. Gertbakh, I. B., and Kh. B. Kordonskiy, *Models of Failure*, Springer-Veriag, New York, 1969.

4. Saeks, R., Singh, S. P., and R. W. Liu, "Fault Isolation via Components Simulation," IEEE Trans. on Circuit Theory, Vol. CT-19, pp. 634-640, 1972.

5. Ransom, M. N., and R. Saeks, "Fault Isolation with Insufficient Measurements," IEEE Trans. on Circuit Theory, Vol. CT-20, pp. 416-417, 1973.

6. Ransom, M. N., and R. Saeks, "A Functional Approach to Fault Analysis,' in this work, New York.

7. Saeks, R., and M. N. Ransom, "Fault Isolation via Term Expansion," Proc. of the 3rd Pittsburgh Symp. on Modelling and Simulation, pp. 224-227, 1973.

8. Ransom, M. N., Ph.D. Dissertation, University of Notre Dame, Notre Dame, Ind., 1973.

9. Buhlmann, H., *Mathematical Methods in Risk Theory*, Springer-Verlag, Heidelberg, 1970.

10. Bailey, A. L., "A Generalized Theory of Credibility," Proc. of the Causalty Actuarial Society, Vol. 37, pp. 7-23, 1945.

11. Tung, L., M.S. Thesis, Texas Tech University, Lubbock, Texas, (in preparation).

12. Luenberger, D. G., *Optimization by Vector Space Methods*, J. Wiley and Sons, New York, 1969.

SYMBOLIC FAULT DIAGNOSIS TECHNIQUES

N. N. Puri

Introduction

In this paper an attempt has been made to present the state of the art in fault diagnosis technique for lumped parameter networks. This problem is very closely tied to the various techniques of finding symbolic network functions, given the topology. Fault isolation essentially involves the determination of the element values of the various components of the faulted network from some experimental data. Experimental data is usually the frequency response, either at some selected frequencies, or at a large number of frequencies a small increment apart. The unfaulted network description should be obtained in either the symbolic state variable form or the Symbolic Transfer Function form, so that given either one, we can obtain the frequency response of the unfaulted network and compare it with the actual faulted system response. The state variable formulation is very instructive and useful if the equipment under consideration is multi-input, multi-output. But for a fewer-input, fewer-output it generates too much information requiring large computer capacity and thereby favoring the Transfer Function approach. The basic approach presented in this paper can be divided into two major phases: symbolic description of the network and element value optimization.

Symbolic Description of the Network.

By this we essentially mean that given the type of network component and their topological interconnection which may be transcribed on a set of input cards (whose format will be discussed in the body of the paper) a computer routine is developed for obtaining the Transfer Function (or the state variable model) in which the complex frequency s, the capacitors, the inductors, the resistors, and the active components appear in their symbolic form. There are various programs already available in the literature [2-6]. The computer routine developed here is different than existing ones and is much simpler and less computer time consuming than the ones based upon Mason's rules [2] or Coate's formula [3] which require determining various order closed-loops. Other methods require numerating all the "Trees" [4] of the

N. N. Puri is a Professor at Rutgers University, New Brunswick, New Jersey and a Consultant to ECOM, Ft. Monmouth, New Jersey.

network graph. The new method used to develop Symbolic Transfer Functions is based upon a method of exterior algebra invented by H. Grassman in 1862 in the classical work called "Ausdehnungslehre" [1].

Element Value Optimization.

This phase accepts the symbolic description, the experimental frequency response of the faulted network and some initial guess of the component values as input and after performing a number of iterations, yields the faulted component values. In this iterative algorithm, called the "conjugate gradient optimization," some functional of the error between the frequency response of the so called model (based upon initial guess of the parameters) and the experimental frequency response is reduced to a minimum.

Feedback Concept

Before we discuss the details of how the Transfer Function of a network is obtained, let us discuss a concept called "Feedback" or "Closed Loop Signal Flow Graph" which reduces the problem of determining both the numerator and the denominator polynominal of a Transfer Function to that of one rule requiring the determination of one determinant polynominal. Consider a Transfer Function F(s).

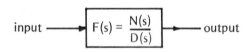

Figure I,a.

Let us modify the above by means of a negative feedback

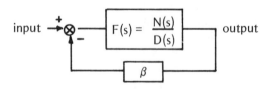

Figure I.b.

The closed loop Transfer Function H(s), of the modified system is given by:

$$H(s) = \frac{F(s)}{1 + \beta F(s)} = \frac{N(s)}{D(s) + \beta N(s)} \qquad (1)$$

Thus if β is coded as a symbol and the denominator $\Delta(s)$ of the closed loop system is determined, then the terms not multiplied by β will belong to D(s) and the terms multiplied by β can be assigned to N(s). Thus it is only necessary to find the denominator of the modified "closed loop" system. The denominator is often referred to as the determinant of the Signal Flow Graph SFG. Details of how SFG can be formulated are given in references [7], [8], and [9]. We shall only refer to some important features and steps in SFG formulation.

Symbolic Network Modeling

Most of the methods used in Symbolic Transfer Function determination involve formulation of a Signal Flow Graph (SFG) which is just a weighted directed graph represented by a system of simultaneous linear equations. The various phases of symbolic modeling can be described as:

a. Input Data Description.
b. Formulation of Signal Flow Graph
c. Implementation of Computer Algorithm chosen to obtain Symbolic Transfer Function.

Input Data Description.

This involves coding the network topology showing how various components are interconnected. Input cards are prepared based upon the topological layout. The following steps should be performed in preparing these cards:

a. Each node, each branch should be numbered.
b. Each branch should be given a symbol. If some branches are considered as unfaulted, their numerical value in appropriate units should be entered.
c. Voltage Sources (independent as well as controlled) should be described first so that they may be selected as the part of a bushy tree to be selected for SFG.
d. Card 1 should form the title of the problem
e. Card 2 is made up of various fields which are right or left adjusted. They describe the number of nodes, number of branches, code base used for symbols and any information that should be printed.
f. Card three identifies the input and output terminals. Various fields record the branch numbers of the input and the output and the nodes identifying the positive side.
g. The cards four to card (B + 3), where B is the number of branches, describe the element type (such as capacitance C, current controlled current source CC etc), element number of the branch, its relative positive and negative node, symbol, its numerical value and the format.

h. This card (B + 4) indicates whether this routine is used to solve more problems or should terminate computation.

Signal Flow Graph SFG Formulation.

Tree Selection Algorithm

Central to SFG is the choice of a network tree. As indicated earlier the voltage elements are the first ones to be used. Then comes the passive elements. In selecting a tree involving a n node network we are essentially selecting a undirected sequence of (n - 1) connected branches forming no closed loop or loops. Suppose a connected sequence of i branches graph forming no loops is selected. The $(i + 1)^{th}$ branch is so selected that:

a. It is connected to the i branch sequence discussed before. This is called the incident node.
b. No path exists between the $(i + 1)^{th}$ branch and the i branches graph.
c. In the case of the bushiest tree, the terminal node of the $(i + 1)$th branch is connected to the largest number of other nodes consistent with (b).

Fundamental Loops or Circuits

The elements of a tree are referred to as branches. The rest of the elements of the networks are called links. Each link l_k forms a fundamental circuit or a loop C_k. The graph consists of tree voltages nodes, and link current nodes. The following set of rules helps to formulate the SFG.

Rule 1. Set $b_{k1}, b_{k2}, \ldots, b_{km}$ form the branches of a tree which form a circuit with link l_k. A set of nodes $V_{bkl}, \ldots, V_{bkm}, l_k$ are established (if they are not already established). For each passive element l_k, a directed graph branch is formed in SFG, directed from V_{bki} ($b_{ki} \neq l_k$) to the node l_k

having weight equal to the admittance of b_{ki} with proper sign, negative when b_{ki} and l_k have same direction, positive otherwise (See Fig. 2).

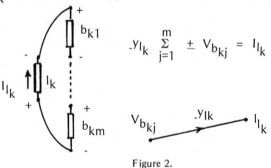

$$-y_{l_k} \sum_{j=1}^{m} \pm V_{b_{kj}} = l_{l_k}$$

Figure 2.

Rule 2. Fundamental Cut Set Equations. Find all the elements of the graph connected to the negative end of the branch b_i (passive) of the selected tree, forming the circuits. All these elements are links. Let us call them $l_{i1}, l_{i2}, \ldots, l_{ij}$.

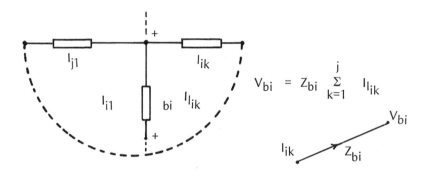

$$V_{bi} = Z_{bi} \sum_{k=1}^{j} I_{l_{ik}}$$

Figure 3.

Thus for each passive link I_k which connected to the branch b_i, a directed SFG graph branch is formed from node $I_{l_{ik}}$ to node V_{bj} with weight equal to the impedance of the branch bi, the weight being positive if the directions of I_{ik} and bi are same, negative otherwise. Thus every time a circuit is formed by a link I_k, the interconnections between the nodes of the voltage branches of the tree and current node of the link are performed.

Rule 3. If any of the four types of controlled sources are present, such that a link or branch of a tree represents controlled sources, an extra controlling variable is created and connected to the rest of the graph though constant of proportionability, impedance or admittance weights, properly directed.

Rule 4. In order to take advantage of the feedback concept discussed in Section 2, the SFG is closed by a symbolic weight β. It is understood that all the weights of the branches of the SFG have either numerical values or symbolic codes. This results in the construction of Signal Flow Graph (SFG). A point of caution should be made regarding various graphs. The Coates flow graph is different than the SFG discussed above. The topological directed graph for determining the "Trees" is different than both of the above.

Transfer Function Evaluation From SFG

An n node signal flow graph represents a system of n independent algebraic equations in n unknowns.

$$\sum_{j=1}^{n} a_{ij}x_j = b_i u \quad (i = 1, \ldots, n) \tag{2}$$

where u represents some independent source. If u is the input and x_j is the output.

In matrix form (2) is given as

$$\underline{A}\,\underline{x} = \underline{b}\,u \tag{3}$$

The Transfer Function is given by

$$F = \frac{x_j}{u} = \sum_{k=1}^{n} \frac{b_k\,(\Delta\, kj)}{\Delta} = \frac{N}{D} \tag{4}$$

where Δ and Δ_{kj} are determinants of $\underline{A}$ and cofactors of a_{kj} respectively. New feedback equations are created as

$$x_{n+1} - x_j = 0$$

$$u_1 - x_{n+1} = u \tag{5}$$

The closed loop equations are

$$\begin{bmatrix} \underline{A} & \underline{b} \\ -\beta \underline{1}_j^t & 1 \end{bmatrix} \begin{bmatrix} \underline{x} \\ x_{n+1} \end{bmatrix} = \begin{bmatrix} \underline{b}\,u_1 \\ 0 \end{bmatrix} \qquad \underline{1}_j = \begin{bmatrix} 0 \\ 0 \\ \vdots \\ 0 \\ 1 \\ 0 \\ \vdots \\ 0 \end{bmatrix} j \tag{6}$$

Determinant of (6) is

$$\Delta_c = \Delta + \beta N \tag{7}$$

Mason's Rule

The Mason's rule for finding the determinant Δ_c can be stated as

$$\Delta_c = 1 + \sum_k (-1)^k \sum_j P_{kj} \tag{8}$$

where

Δ_c = closed loop graph determinant.

P_{kj} = a product of loop gains of the jth set of k non-touching loops.

P_{kj} = sum of all the kth order loops, or the sum of all the k non-touch loops.

The essential steps involved in the Mason's rules are
  a.   The loop finding algorithm.
  b.   The determination of the kth order non-touching loops.
  c.   A coding scheme to store the terms involved in the product of loop gains.

Node Elimination Technique

This technique can be used either by itself to find the Transfer Function or used to simplify SFG so that the Mason's rule can be applied to solve some reasonable size network. The SFG is made up of the input node (single input), output node (single output) and the secondary nodes x_k, $k = k_1$, $k_2, \ldots, k_n$. Let

  $x(i,k)$ = The nodes which are incident upon x_k, called the incidence nodes. This means the signal flows from $x(i,k)$ to x_k via the branch $g(i,k),(i = i_1, i_2 \ldots, i_m)$.

  $x(k,t)$ = Nodes upon which the node x_k terminates, called the target nodes. This means the signal flows from x_k towards $x(k,t)$ via the branch $g(k,t)$.

The SFG is so formulated that none of the nodes have self loops. The situation with self loops is quite easily modified but the symbolic manipulation becomes quite tedious.

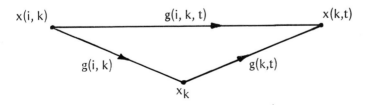

$g(i, k, t)$ = Transmittance between incident node $x(i,k)$ and target node properly directed.

After the node x_k is eliminated the transmittance $g(i, k, t)$ [1] is modified as

$$g(i, k, t) = g(i, k, t) + g(i, k) g(k, t) \qquad (9)$$

$x(i,k)$ •————————————▶———————————• $x(k,t)$

$$g(i, k, t) = g(i, k, t) + g(i, k) g(k, t) (1-g_{kk})^{-1}$$

This elimination process is performed for all the secondary nodes till only the input and the output nodes are left. In case the node x_k is allowed a self loop with transmittance g_{kk}, the term $g(i, k) g(k, t)$ is multiplied with $(1 - g_{kk})$-1. The resultant transmittance between the input node and the output node, when all the secondary nodes are eliminated yields the final Transfer Function.

Coates Formula

To the set of algebraic equations

$$\sum_{j=1}^{n} a_{ki}x_j - b_k u = 0 \quad (k = 1, 2, \ldots n) \qquad (10)$$

A flow graph FG (not SFG) is associated. It has one source node (called node o) and n variable nodes. To the source node is associated an input variable u and to other nodes is associated one of the variables $x_1, x_2, \ldots x_n$. For $a_{kj} \neq o$, a branch of weight a_{kj} is formed directed from node j to node k. For each $b_k \neq o$ a branch of weight $-b_k$ is formed from node o to node k. The equation at a node k is obtained by equating to zero (and not to x_k as in the Mason's SFG) the sum of product of their branch weight times the variables, these branches originating from and terminating on x_k. The concept of feedback can be applied by letting the output variable node x_j connected to source node through $-\beta$.

The Transfer Function is

$$T = \frac{x_j}{u} = \frac{N}{D}$$

In order to find both N and D from the same formula, a closed loop system is created by a new extra node x_{n+1} and modifying the existing graph as shown by the solid lines.

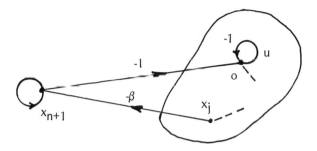

The new graph has n + 1 nodes and is called G(n+1).

A connection of the flow graph G(n+1) is a subgraph of all its nodes such that only one branch originates and only one branch terminates on each node. The modified Coates formula can now be stated as

$$\Delta c = D + \beta N = (-1)^{n+2} \sum_k (-1)^{l_k} C(G(n+1))_k \tag{11}$$

where

    n = The number of dependent variables in the system of simultaneous equations.

    l_k = The number of directed loops in the kth connection.

$((G(n+1))_k$ = The connection gain of the kth connection. The summation is taken over all the connections of G(n+1).

    This formula is supposed to give fewer cancellations than the Mason's formula.

The Tree Enumeration Method

    The basic concepts of the topological analysis are not new. The original papers of Kirchoff and Maxwell date back to mid 18th century [10]. See also the papers by Percival [10] and Kuh [15]. The approach is based upon Binet-Cautchy Theorem [13, 14] which evaluates the determinant of the product of two rectangular (n x m) and (m x n) matrices in terms of the major determinants of individual matrices.

    Consider a passive network with b branches and $\underline{Y}_b$ a b x b diagonal admittance matrix. If $\underline{A}$ is defined as the reduced incidence matrix with (n - 1) nodes and b edges, showing the interconnection of nodes and branches, then the expression for the determinant (or c) with proper modification can be given as

$$\Delta = |\underline{A}\ \underline{Y}_b\ \underline{A}^t\ | = \Sigma \text{ tree admittance product of networks} \tag{12}$$

Thus all the trees of the graph have to be determined.

The method has been extended to active networks. In this case two directed graphs DG (not the SFG or FG) called the voltage and current graphs are produced. According to Percival [10] consider a pair of passive node (r,t) and voltage-current relation

$$Yrt \ Vrt = Irt \tag{13}$$

The directed graph is shown below

$$\underset{\bullet \hspace{3.5cm} \bullet}{\overset{r \hspace{1.3cm} Y_{rt} \hspace{1.3cm} t}{\rule{3.5cm}{0.4pt}}}$$

Now consider joint pair pg and mn such that

$$g_m \ V_{mn} = I_{pg} \tag{14}$$

A joint voltage current dual graph is given as

The solid arrow represents the voltage edge and double arrow represents the current edge. A current graph G_I is formed consisting of all the passive edges and all the active current edges with two arrows. Similarly a voltage graph G_v is formed consisting of all the passive edges and voltage edges formed by the solid black arrow. The determinant Δ is now obtained as $\Delta = \Sigma \pm$ (tree products of the common trees of the G_I and G_v of the network) (15). This complicates the tree finding procedure. The procedure is time consuming. Similar rules can be found for impedance also. For passive networks only the sum of all the tree products are needed to yield Δ.

Exterior Algebra

Consider an n dimensional vector space V_n over a field R of real numbers a, b, ... Let $\alpha, \beta \ldots$, be the elements of the space. We call $\underline{\alpha} \wedge \underline{\beta}$ to be the outer (or exterior) products of $\underline{\alpha}$ and $\underline{\beta}$ subject to following rules:

$$(a_1 \alpha_1 + a_2 \alpha_2) \wedge \underline{\beta} - a_1 (\alpha_1 \wedge \underline{\beta}) - a_2 (\alpha_2 \wedge \underline{\beta}) = \underline{0}$$

$$\underline{\alpha} \wedge (b_1 \alpha_1 + b_2) - b_1 (\alpha \wedge \alpha_1) - b_2 (\alpha_2 \wedge \alpha_2) = \underline{0}$$

$$\underline{\alpha} \wedge \underline{\alpha} = 0 \tag{15}$$

$$\underline{\alpha} \wedge \beta + \beta \wedge \alpha = 0$$

If $\underline{\alpha}$ and $\underline{\beta}$ are dependent, say $\underline{\beta} = c\,\underline{\alpha}$

Then $\underline{\alpha} \wedge \underline{\beta} = \underline{\alpha} \wedge (c\,\underline{\alpha}) = c \cdot \underline{0} = \underline{0}$

Suppose $e_1, e_2, \ldots, e_n$ is a basis of the space L.

Let

$$\underline{\alpha} = \sum_{i=1}^{n} a_i\,\underline{e}_i$$

$$\underline{\beta} = \sum_{i=1}^{n} b_i\,\underline{e}_i$$

$$\underline{\alpha} \wedge \underline{\beta} = (\sum_{i=1}^{n} a_i\,\underline{e}_i) \wedge (\sum_{j=1}^{n} b_j\,\underline{e}_j) = \sum_{i=1}^{n} \sum_{j=1}^{n} a_i\,b_i\,(\underline{e}_i \wedge \underline{e}_j)$$

Considering

$$\underline{e}_j \wedge \underline{e}_i = -\underline{e}_i \wedge \underline{e}_j, \quad \underline{e}_i \wedge \underline{e}_i = 0$$

The exterior product is given as

$$\underline{\alpha} \wedge \underline{\beta} = \sum_{i<j} (a_i\,b_j - a_j\,b_i)\,\underline{e}_i \wedge \underline{e}_j \tag{16}$$

Application of Exterior Algebra to Transfer Function Evaluation

Let us write the equations representing the signal flow graph as

$$a_{11}\,x_1 + a_{12}\,x_2 + \cdots + a_{1n}\,x_n = b_1 u$$

$$a_{21}\,x_1 + a_{22}\,x_2 + \cdots + a_{2n}\,x_n = b_2 u \tag{17}$$

$$a_{n1}\,x_1 + a_{n2}\,x_2 + \cdots + a_{nn}\,x_u = b_n u$$

Multiplying the equations with $\underline{e}_1, \underline{e}_2, \ldots, \underline{e}_n$ respectively and adding them together yields a vector equation

$$\underline{a}_1\,x_1 + \underline{a}_2\,x_2 + \cdots + \underline{a}_n\,x_n = bu \tag{18}$$

Where $\underline{a}_1, \underline{a}_2, \ldots, \underline{b}$ are vectors in the spave v_n defined before. This space is the exterior product space where

$$\underline{a}_k = a_{1k}\,\underline{e}_1 + a_{2k}\,\underline{e}_2 + \cdots + a_{nk}\underline{e}_n \tag{19}$$

The Transfer Function

$$T = \frac{xj}{u} = \frac{a_1 \wedge a_2 \wedge \cdots \wedge a_{j-1} \wedge b \wedge a_{j+1} \cdots \wedge a_n}{a_1 \wedge a_2 \wedge a_3 \cdots \wedge a_n} \tag{20}$$

Thus the determinant of the graph is given by the exterior product

$$\Delta = a_1 \wedge a_2 \wedge a_3 \wedge \cdots \wedge a_n \tag{21}$$

As stated before the graph can be augmented and the determinant of the closed loop system can give both the numerator and the denominator. The expression (21) is very useful and informative in the sense that all the vectors need not be multiplied together at once.

Computer Determination of Transfer Function Via Exterior Algebra

Determinant Evaluation

Example 1

We shall show the computational procedure by means of an example. Let us consider a signal graph which yields the set of equations

$$\underline{A}\ \underline{x}\ =\ \underline{b}\ \underline{u} \tag{22}$$

where

$$\underline{A} = \begin{bmatrix} a_{11} & a_{12} & a_{13} & 0 \\ 0 & a_{22} & 0 & a_{24} \\ a_{31} & 0 & a_{33} & 0 \\ 0 & a_{42} & a_{43} & a_{44} \end{bmatrix} \qquad \underline{b} = \begin{bmatrix} 1 \\ 0 \\ 0 \\ 0 \end{bmatrix}$$

$$\tag{23}$$

Let us obtain the determinant using following steps.

Corresponding to each column of matrix $\underline{A}$ form vectors $\underline{a}_1$, $\underline{a}_2$, etc., such that the row of the non zero element are listed as entries of the vectors. Thus:

$$\underline{a}_1 = (1, 3)$$
$$\underline{a}_2 = (1, 2, 4)$$
$$\underline{a}_3 = (1, 3, 4) \tag{24}$$
$$\underline{a}_4 = (2, 4)$$

Corresponding to $\underline{a}_1, \underline{a}_2, \underline{a}_3$, and $\underline{a}_4$ with

$\Delta_1 = 1 + 3$

$\Delta_2 = 1 + 2 + 4$

$\Delta_3 = 1 + 3 + 4$

$\Delta_4 = 2 + 4$

Now let

$\Delta^{(1)} = \Delta_1 = 1 + 3$

$\Delta^{(2)} = \Delta^{(1)} \Delta_2 = (1 + 3)(1 + 2 + 4)$

(25)

The rules of multiplication are the same as ordinary multiplication, except when the two letters (which stand for symbols) are repeated, the term becomes zero. For example: $11 = 0, 22 = 0$, etc.

Thus

$\Delta^{(2)} = 12 + 14 + 31 + 32 + 34$

$\Delta^{(3)} = \Delta^{(2)} \Delta_3$

$= (12 + 14 + 31 + 32 + 34)(1 + 3 + 4)$

Once again terms such as 121, 141 with repeated symbols cancel and we obtain

$\Delta^{(3)} = (32 + 34)(1) + (12 + 14)(3)$

$\quad + (12 + 31 + 32) 4$

$= 321 + 341 + 123 + 143 + 124 + 314 + 324$

Similarly

$\Delta^{(4)} = \Delta = \Delta^{(3)} \Delta_4$

$= (321 + 341 + 123 + 143 + 314 + 324)(2 + 4)$

$= 3412 + 1432 + 3142 + 3214 + 1234$

(26)

The order of multiplication of symbols in each term of Δ is preserved. Let us consider a typical term 3142. Letter 3 in the first position stands for the element of the first column and third row. Letter 1 in the second position stands for the element of the second column and first row. Thus,

$$3 \quad 1 \quad 4 \quad 2 \longleftrightarrow a_{31} \quad a_{12} \quad a_{43} \quad a_{24}$$

We have to determine the proper signs (plus or minus) associated with 3142. This is done as following:

Consider the sequence 3142.

After one interchange of adjacent symbols, 1 is brought to the first position. Let $S^{(1)} = 1$.

Remove 1 from the sequence 3142. The remaining sequence is 342. It takes two interchanges of adjacent symbols to bring 2 to the first position. Thus $S(2) = 2 + S(1) = 3$.

Remove 2 from the sequence 34. This sequence is in proper order. Thus $S = S(2) = 3$. Hence

$$3142 = (-1)^3\ a_{31}\ \ \ a_{12}\ \ \ a_{43}\ \ \ a_{24}$$

Thus the determinant Δ is given as

$$\Delta = (-1)^4 a_{31} a_{42} a_{13} a_{24} + (-1)^3 a_{11} a_{42} a_{33} a_{24}$$
$$+ (-1)^3\ a_{31} a_{12} a_{43} a_{24} + (-1)^3 a_{13} a_{22} a_{31} a_{44} + a_{11} a_{22} a_{33} a_{44}$$

$$(27)$$

It should be noted that there is no cancellation of the terms.

Numerator Evaluation (addition of extra node)

Let us assume that for the example discussed before we are required to find the Transfer Function.

$$T = \frac{x_4}{u} = \frac{N}{D}$$

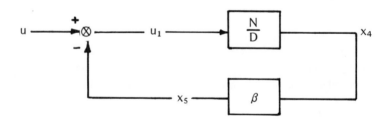

Let us form a closed loop system as shown

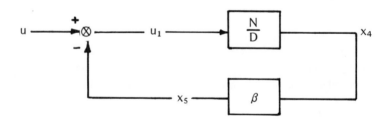

The new equations are

$$\underline{A}\ \underline{x} = bu_1 = b\,(u\text{-}x_5)$$
$$\beta x_4 - x_5 = 0$$

$$(28)$$

The closed loop matrix is given as

$$\underline{A}_c = \left[\begin{array}{ccc|c} & & & b \\ & & & 0 \\ & \underline{A} & & 0 \\ & & & 0 \\ \hline 0 \;\; 0 \;\; 0 \;\; \beta & & & -1 \end{array} \right] \tag{29}$$

Thus $\underline{a}_1, \underline{a}_2, \underline{a}_3$, remain unchanged.
Thus a_1, a_2, a_3, remain unchanged.
$\underline{a}_4 = (2, 4, 5)$, 5 represents the feedback.
$\underline{a}_5 = (1, 5)$

To find the numerator terms only we form a new vector $\underline{a}_4$ which contains feedback only. Thus,

$$\underline{a}_4' = (5)$$

$$\Delta_4 \Delta_5 = 5 \, (1 + 5) = 51$$

$$\begin{aligned} N = [\, \Delta(5)]' &= \Delta(3) \; \Delta_4' \; \Delta_5 \\ &= (321 + 341 + 123 + 143 + 124 + 314 + 324) \, (51) \\ &= 32451 \end{aligned}$$

Thus

$$N = (-1)^{4+1} \, a_{31} \; a_{22} \; a_{43} (1) \; (b) = a_{31} \; a_{22} \; a_{43} \; b \tag{30}$$

Transfer Function Evaluation with No Extra Node

Example 2

Consider a set of equations

$$\begin{aligned} a_{11} \, x_1 + a_{12} \, x_2 + a_{13} \, x_3 &= b_1 \\ a_{21} \, x_1 + a_{22} \, x_2 + a_{23} \, x_3 &= 0 \\ a_{31} \, x_1 + a_{32} \, x_2 + a_{33} \, x_3 &= 0 \end{aligned} \tag{31}$$

we are required to find x_1

We set the following new equations

$$a_{11}\, x_1 + a_{12}\, x_2 + a_{13}\, x_3 = b_1 = b - x_4 = b - \beta\, x_1$$
$$a_{21}\, x_1 + a_{22}\, x_2 + a_{23}\, x_3 = 0 \tag{32}$$
$$a_{31}\, x_1 + a_{32}\, x_2 + a_{33}\, x_3 = 0$$

Thus

$$\underline{a}_1 = (1,2,3)$$
$$\underline{a}_2 = (1,2,3)$$
$$\underline{a}_3 = (1,2,3)$$

The terms in $\Delta^{(3)}$ which correspond to 1 as the first integer can be identified as the numerator terms. Thus,

$$\Delta^{(1)} = (1 + 2 + 3)$$
$$\Delta^{(2)} = \Delta^{(1)} \Delta_2 = (1 + 2 + 3)\,(1 + 2 + 3)$$
$$= (12 + 13 + 21 + 23 + 31 + 32)$$
$$\Delta^{(3)} = \Delta^{(2)} \Delta_3 = (12 + 13 + 21 + 23 + 31 + 32)\,(1 + 2 + 3)$$

Thus

$$\Delta = 123 + 132 + 213 + 231 + 312 + 321$$
$$N = (23 + 32)$$

Thus

$$D = \Delta = \quad a_{11}\, a_{22}\, a_{33} + (\text{-}1)\, a_{11}\, a_{32}\, a_{23} + \quad (\text{-}1)\, a_{21}\, a_{12}\, a_{33} \tag{33}$$
$$+ (\text{-}1)^2\, a_{21}\, a_{32}\, a_{13} + (\text{-}1)^2\, a_{31}\, a_{12}\, a_{23} + (\text{-}1)^3\, a_{31}$$
$$a_{22}\, a_{13}$$

$$N = a_{22}\, a_{33} - a_{32}\, a_{23} \tag{34}$$

In order that Exterior Algebra can be an effective tool, the following points should be observed.

The SFG should be formed from a "Bushy Tree" resulting in a favorable $\underline{A}$ matrix.

The columns of $\underline{A}$ should be so arranged that if $m_1, m_2, \ldots, m_n$ are the number of entries in the first, second, . . . , nth column, then

$$m_1 < m_2 < m_3 < \ldots < m_n$$

Symbolic Coding of the Elements of the Network

Each element of the network and hence the branch of the SFG is given a weight $w(b)$. The branch weight is given as

$$w(b) = k \cdot x \cdot s^n$$
$$k \quad = \text{constant}$$
$$x \quad = \text{symbol} \tag{35}$$
$$s \quad = \text{complex frequency}$$

Thus three parameters k, x, and n define the branch weight w (b). The symbol x represents the particular resistor, capacitor or inductor and is coded by module two base number B, chosen from the set $(2^1, 2^2, 2^3, \ldots, 2^m)$. Thus the symbols are coded as

Symbol	Code Integer		Code Integer
x_1			B^o
x_2			B^1
$\vdots$			$\vdots$
x_k			B^{k-1}

The base number B is so chosen that at the most (B-1) branches of SFG have the same designation [2]. In our case all branches are distinct and hence B = 2. This essentially means that in any of the terms of N or D any symbol x will appear at the most linearly [2]. Thus decoding of the terms will be unique if B = 2 is used.

Thus consider a term $\alpha \, C^3 \, R$.

Highest Power of a Symbol = 3 = B - 1

B = 4

Let

$$R = B^0 = 4^0 = 000001$$
$$C = B^1 = 4^1 = 000100$$
$$\alpha = B^2 = 4^2 = 010000$$

$$\alpha \, C^3 \, R = \underline{01} \ \underline{11} \ \underline{01}$$

Decoding of the module 2 number is obvious.

Element Value Optimization

Consider the Transfer Function of the unfaulted network to be available in the form

$$F(s) = \frac{N(s)}{D(s)} = \frac{\sum\limits_{i=1}^{m} (b_i) \ s^{m-i}}{s^n - \sum\limits_{i=1}^{n} a_i \ s^{n-1}} \qquad (36)$$

where

$$a_i = f_i(x_1, x_2, \ldots, x_r) \qquad (i = 1, \ldots, m)$$
$$b_k = g_k(x_1, x_2, \ldots, x_r) \qquad (k = 1, \ldots, n) \qquad (37)$$

Where f_i and g_k represent non-linear relations between various components of the network designated by the symbols $x_1, x_2, \ldots, x_r$. Consider also that the measured frequency response of the given faulted as $M(jw)$ where $M(jw)$ is a complex number for $|w| \leq w_0$ such that $|M(jw) M(0)^{-1}| \ll 1$. The section develops the conjugate gradient technique to obtain the numerical values for the symbols $x_1, x_2, \ldots, x_r$.

Let us represent $F(s) = F(s, \underline{x})$ to emphasize the fact that $F(s)$ is a function of variables $\underline{x}$. The fault isolation algorithm can now be defined as the choice of the vector $\underline{x}$ in such a manner as to minimize

$$
\begin{aligned}
J(\underline{x}) &= \int_0^{w_0} \left(\frac{F(jw, \underline{x})}{M(jw)} - 1 \right) \left(\frac{F(jw, \underline{x})}{M(jw)} - 1 \right)^* dw \\
&= \int_0^{w_0} Y(jw, \underline{x}) \, dw
\end{aligned}
\tag{38}
$$

The symbol "*" stands for complex conjugate.
The crux of the algorithm involves determining the gradient $\partial J / \partial x_i, \, i = 1, \ldots, r.$

It is evident from (38) that

$$
\begin{aligned}
\frac{\partial J}{\partial x_i} &= \int_0^{w_0} \left[\frac{\partial F}{\partial x_i} \frac{(F-M)^*}{MM^*} + \left(\frac{F-M}{MM^*} \right) \frac{\partial F^*}{\partial x_i} \right] dw \\
&= \text{Twice the real part of } \left\langle \frac{\partial F}{\partial x_i} , \frac{F-M}{MM^*} \right\rangle
\end{aligned}
$$

$$(i = 1, \ldots, r) \tag{39}$$

The symbol $\langle , \rangle$ stands for inner product in Hilbert Space

Evaluation of the Gradient

From chain rule

$$
\frac{\partial F}{\partial a_i} = \sum_{i=1}^{n} \frac{\partial F}{\partial a_i} \frac{\partial a_i}{\partial x_i} + \sum_{k=1}^{m} \frac{\partial F}{\partial b_k} \frac{\partial b_x}{\partial x_i}
\tag{40}
$$

$$
\frac{\partial F}{\partial a_i} = -F(jw)(jw)^{n-i}(D(jw))^{-1}
$$

$$
\frac{\partial F}{\partial b_k} = F(jw)(jw)^{m-k}[N(jw)]^{-1}
\tag{41}
$$

We shall now summarize the total algorithm for fault isolation in the network.

1. From the wiring diagram prepare the input cards and code various elements.

2. Select a bushy tree and draw an SFG.

3. Using any of the topological methods finds the Transfer Function in Symbolic form.

4. Find the relationship between various $x_1, x_2, \ldots, x_r$ and the coefficients a_i and b_k of the denominator and the numerator dynamics.

5. Initialize by choosing x_j $(j = 1, \ldots, r)$ and symbolize it as $x_j^{(0)}$. Compute $a_i^{(0)}, b_i^{(0)}, F(0)$ and partial derivatives of F.

6. Select a frequency range w_o and incremental frequency Δw. Store the real and imaginary parts of $M(jw)$ for $w = 0, \Delta w, \ldots, w_o$.

7. Compute

$$J(\underline{x}_o) \, , \, \underline{g}^{(0)} = \underline{\nabla} \times (J)^o = \begin{bmatrix} \dfrac{\partial J}{\partial x_1}(o) \\ \vdots \\ \dfrac{\partial J}{\partial x_r}(o) \end{bmatrix} \qquad (42)$$

This involves using (40) and (41).

8. Let

$$\underline{s}^{(0)} = -\underline{g}^{(0)} \, , \text{ the initial search direction} \qquad (43)$$

9. $\underline{x} = \underline{x}^{(i)} + \alpha \, \underline{s}^{(i)}$

Compute $\alpha = \alpha^1$ for which the value of the functional

$J^{(i)}(\alpha) = J(\underline{x}^{(i)} + \alpha \, \underline{s}^{(i)})$ is minimized. This is performed by choosing $\Delta\alpha$ such that

$$\alpha = k\Delta\alpha \ (k = 1, 2, \ldots)$$

At each increment the quantity $d \, J^{(i)}(\alpha)$ is computed and its sign observed. $\quad d\alpha$

$$\frac{d}{d\alpha} \, J^{(i)}(\alpha)\big|_{\alpha = k\Delta\alpha} = \langle \underline{g}_k^{(i)} , \, \underline{s}^{(i)} \rangle$$

where

$$g_k^{(i)} = \underline{\nabla}_x \, J(\underline{x}) \big|_{\underline{x} = \underline{x}^i + K\Delta\alpha \underline{s}^i}$$

The quantity α is continued to be incremented till the sign $\dfrac{(d)}{(d\alpha)}$ $J^i(\alpha)$ is reversed. Let α_1 and α_2 be the last two values of α such that $\dfrac{(d)}{(d\alpha)} J^i(\alpha)$ changes sign. A cubic interpolation may be applied between the two points α_2 and α_1 to obtain the α at which the minimum of $J^{(i)}(\alpha)$ is obtained. The results of the cubic interpolation can be summarized as follows:

$$\frac{d}{d\alpha} J^i(\alpha) = f^{(i)}(\alpha) \qquad\qquad \frac{d^2}{d\alpha^2} J^i(\alpha) = f'^{(i)}(\alpha)$$

$$Z = f'^{(i)}(\alpha_1) + f'^{(i)}(\alpha_2) + \frac{3}{\alpha_2 - \alpha_1}\left(f^{(i)}(\alpha_1) - f^{(i)}(\alpha_2)\right)$$

$$w = |[z - f'^{(i)}(\alpha_1)\, f'^{(i)}(\alpha_2)]^{1/2}|$$

$$\alpha^{(i)} = \alpha_2 - (\alpha_2 - \alpha_1)\frac{2 + f'^{(i)}(\alpha_1) \pm w}{2Z + f'^{(i)}(\alpha_1) + f'^{(i)}(\alpha_2)}$$

The "$\pm$" ambiguity is resolved from the fact that

$$\alpha_1 \leqslant \alpha^{(i)} \leqslant \alpha_2$$

10.   Update the $\underline{x}$ vector as

$$\underline{x}^{(i+1)} = \underline{x}^{(i)} + \alpha^{(i)}\,\underline{s}^{(i)}$$

11.   Compute the new gradient

$$\underline{g}^{(i+1)} = \nabla \times J\,(\underline{x})\Big|_{\underline{x}\,=\,\underline{x}^{(i+1)}}$$

$$\beta^{(i)} = \frac{\langle \underline{g}^{(i+1)}, g^{(l+1)}\rangle}{\langle g^{(i)}, g^{(i)}\rangle}$$

$$\underline{s}^{(i+1)} = -\underline{g}^{(i+1)} + \beta^{(i)}\,\underline{s}^{(i)}$$

REFERENCES

1. Grassman, H., "Die Wissenshaft Der Extensiven Grosse," Erster Theil, die Lineale Ausdehnungslehre xxxii+279 pp. 1844 Leipzig.

2. Lin, P. M., et al, "SNAP — A Computer Program for Generating Symbolic Network Functions," School of Electrical Eng'g., Purdue Univ., Lafayette, Ind. Rep TR-EE70-16, August 1970.

3. Cotes, C. L., "Flow-graph Solutions of Linear Algebraic Equations," IRE Trans. on Circuit Theory, Vol. CT6, pp. 170-187, June 1959.

4. Minty, G. J., "A Simple Algorithm for Listing all the Trees of a Graph," IEEE Trans. on Circuit Theory (correspondence),Vol. CT-12, p. 120, March 1965.

5. Pottle, C., "CORNAP User's Manual, School Elec. Eng., Cornell Univ. Ithaca, N. Y., 1968.

6. McNamee, L. P. and Potach, H., "A Users and Programmers Manual for NASAP," Univ. Calif., Los Angeles, Rep 68-38, Aug. 1968.

7. Lin, P. M., "A Survey of Applications of Symbolic Network Functions," IEEE Trans. Circuit Theory, Vol. CT-20, No. 6, pp. 732-737, Nov. 1973.

8. Calahan, D. A., "Computer Aided Network Design Preliminary Edition, McGraw-Hill Book Co., New York, 1968.

9. Branin, F. H., Jr., "Computer Methods of Network Analysis," Proc. IEEE, Vol. 55, No. 11, November 1967.

10. Percival, W. S., "The Solution of Passive Electrical Networks by Means of Mathematical Trees," Proc. IEE (London) Vol. 100C, pp. 143-150, 1953.

11. Seshu, S. and Waxman, R., "Fault Isolation in Conventional Linear Systems — A Feasibility Study," IEEE Trans. Reliability, Vol. R-15, pp. 11-16, May 1966.

12. Mayeda, W. and Seshu, S., "Generation of Trees Without Duplication," IEEE Trans. Circuit Theory, Vol. CT-12, pp. 181-185, June 1965.

13. Bashkow, T., "The A Matrix, a New Network Description, " IRE Trans. on Circuit Theory, Vol. CT-4 pp. 117-119, September 1957.

14. Bryant, P. R., "The Explicit Form of Bashkow's A Matrix, " IRE Trans. on Circuit Theory, Vol. CT-9 CT-9, pp. 303-306, 1962.

15. Kuh, E. S. and Rohrer, R. A., "The State-Variable Approach to Network Analysis," Proc. IEEE, Vol. 53, pp. 672-686, July 1965.

16. Chen, W. K., "Applied Graph Theory," Amsterdam, The Netherlands, North-Holland 1971, CHE 4.5.

17. Bass, S. C., "The Application of a Fast Symbolic Analysis Routine in a Network Optimization Program," Proc. 15th Midwest Symp. Circuit Theory, 1972.

18. Comer, D. J., "Computer Analysis of Circuits," International Text Book Company, 1971.

19. Sedor, S. R., "SCREPTE: A Program for Automated Network Analysis," IBM Journal, Vol. 11, pp. 627-637, November 1967.

20.   , "1620 Electronic Circuit Analysis Program (ECAP)," (1620-EE-02X) Users Manual, IBM Publication H20-0170-1, IBM Corporation, White Plains, N. Y.

21. Van Landingham, H. F., "A Unified Method for Obtaining the State-Variable Model for Time Variable Networks Containing Controlled Sources," Proc. of the First Asilomar Conference on Circuits and Systems pp. 143-152, 1967.

22. McCalla, W. J. and Pederson, D. O., "Elements of Computer-Aided Circuit Analysis," IEEE Trans. Circuit Theory (Special Issue on Computer Aided Circuit Design), Vol. CT-18, pp. 14-26, January 1971.

23. Desoer, C. A., "The Optimum Formula for the Gain of a Flowgraph or a Simple Derivation of Coates Formula," Proc. IRE, Vol. 48, pp. 883-889, 1960.

24. Happ, W. W., "Signal-flow Graphs," Proc. IRE, Vol. 41, pp. 1144-1156, 1953.

25. Chen, W. K., "Unifold Theory on Topological Analysis of Linear Systems," Proc. IEE, Vol. 114, No. 11, pp. 1630-1636, 1967.

27. Acar, C., "Formulation of the State Equations by Signal-flow Graphs," Elect. Lett., Vol. 6, pp. 82-84, 1970.

27. Nathan, A., "A Two-Step Algorithm for the Reduction of Signal Flow Graphs," Proc. Ins. Radio Eng. Vol. 49, p. 1431, 1961.

28. Nathan, A., "A Proof of the Generalized Topological Kirchoffs Rules," Proc. IEE, Vol. 109, Pt. C, pp. 45-50, March 1962.

29. Chen, W. K., "Topological Analysis for Active Networks," IEEE Trans. on Circuit Theory, Vol. CT-12, pp. 85-91, March 1965.

30. Talbot, A., "Topological Analysis of General Linear Networks," IEEE Trans. on Circuit Theory, Vol. CT-12, pp. 170-180, June 1965.

31. Fisher, G. J. and Wing, O., "Computer Recognition and Extraction of Planar Graphs from the Incidence Matrix," IEEE Trans. on Circuit Theory, Vol. CT-13, pp. 154-163, June 1963.

32. Jacob, J. P., "The Number of Terms in the General Gain Formula for Coates and Mason Signal Flow Graphs," IEEE Trans. on Circuit Theory (correspondence) Vol. CT-12, pp. 601-604, Dec. 1965.

FAULT DETECTION AND LOCATION
IN ANALOG CIRCUITS
A bibliography

J. C. Rault, R. Garzia, and S. D. Bedrosian

1   C. Aamand, "Process Identification for On-Line Optimization." EUROCON71, Palais de Beaulieu, Lausanne, Switzerland, October 1971.

2   R.J. Allen, "Failure Prediction Employing Continuous Monitoring Techniques." IEEE Transactions on Aerospace Support Conference, Proc., 1963.

3   A.V. Balakrishnan, and V. Peterka, "Identification in Automatic Control Systems." *Automatica*, Vol. 5, pp. 817-829, 1969.

4   D.R. Barney, P.K. Giloty, and H.G. Kirnzle, "System Testing and Early Field Operation Experience." *BSTJ*, Vol. 49, December 1970.

5   R.F. Barry, R.S. Fisher, and R.R. Mattison, "Programmed Algorithm for Test Point Selection and Fault Isolation." RCA Aerospace Systems Division, Burlington, Mass., Report on Contract No. AF 33(615)-3033, Vol. 1 and 2, August 1966, Also AD-487 520 and AD-487 521.

6   R.F. Barry, "Fault Isolation by Parameter Identification." Automatic Support Symposium, October 1968.

7   R.F. Barry, "Some Analytic Fault Isolation Techniques." Lecture Notes of Computer Aided Testing and Fault Identification of State Systems, The University of Wisconsin, May 23, 1968.

8   J. Baumeister, "In Flight Analysis of Vehicle Dynamics Along the Trajectory by Correlation Techniques." Astrionics Laboratory Marshall Space Flight Center, Huntsville, Alabama, January 1969.

9   E.J. Bayly and V.S. Leradi, " A Method of Dynamic System Testing."Proceedings of the Eighth Conference on Military Electronics, September 1964.

10 P.W. Becker and J.E. Thamcrup, "Pattern Recognition Applied to Automated Testing." The Automation of Testing, IEE Conference No. 91, September 20-22, 1972, pp. 82-86.

11 C. Beckman, "Study of Piece Part Fault Isolation by Computer Logic." Institute for Cooperative Research, University of Pennsylvania, June 1960, Contract No. DA-36-034-507-ORD-3037 TRD.

12 C. Beckman, J. Adler, S.D. Bedrosian, R.S. Berkowitz, and T.C. Chen, "Study of

J. C. Rault is with Thomson-CSF, DIB, Service Recherche, 33 rue de Vouille, 75015 Paris, France.

R. Garzia is with Babcock and Wilcox Inc., Barberton, Ohio 44203.

S. D. Bedrosian is with University of Pennsylvania, Philadelphia, Pa. 19174.

Piece Part Isolation of Computer Logic." Vol. 4, AD-136 615, June 1964, Vol. 5, AD-615 337, June 1964.

13   S.D. Bedrosian and R.S. Berkowitz, "Solution Procedure for Single-Element-Kind Networks," IRE International Convention Record, Vol. 10, Part 2, 1962, p. 16.

14   S.D. Bedrosian, "A Simplified Explicit Solution of Networks with Two Internal Nodes." IEEE Trans. Communication and Electronics, Vol. 71, p. 219, March 1964.

15   S.D. Bedrosian, "On Element Value Solution of Single-Element-Kind Networks." Ph.D. Dissertation, University of Pennsylvania, December 1961.

16   A. Berg, "A/D and D/A Converter Testing." *Electronic Design*, April 1, 1974.

17   R.S. Berkowitz, "Conditions for Network Element Value Solvability." *IRE Transactions on Circuit Theory*, Vol. CT-9, No., pp. 24-29, March 1962.

18   R.S. Berkovitz and Wexelblatt, "Statistical Considerations in Element Value Solutions," IRE Transactions on Military Electronics, Vol. MIL-6, No. 3, pp. 282-88, July 1962.

19   R.S. Berkowitz and P.S. Krishnaswamy, " Computer Techniques for Solving Electric Circuits for Fault Isolation." IEEE Transactions on Aerospace Support Conference Procedures, Vol. AS-1, No. 2, pp. 1090-1099, August 1963.

20   R.S. Berkowitz, W.G. Faust, and M.M. Vartanian, "Validation of Theoretical Automatic Checkout Techniques." Technical Report, AFAPL-TR-68-120, Oct. 1-295, October 1968.

21   J.G. Bollinger and J.A. Bonesho, "Pulse Testing in Machine Tool Dynamic Analysis." *International Journal Mach. Tool Des. Res.*, Vol. 5, 1965.

22   P.M. Boor and R.L. Grimmer, "A Computerized Data Acquisition System for Real Time Fault Isolation." Lockheed Aircraft Service Co., November 1968.

23   J.M. Brown, D.R. Towill and P.A. Payne, "Predicting Servomechanism Dynamic Errors from Frequency Response Measurements." *Radio and Electronic Engineer*, Vol. 42, No. 1, pp. 7-20 January 1972.

24   J.M. Brown, "Sensitivity and Covariance Matrices for Predicting Variability in Response." UWIST DAG TN 49, February 1972.

25   J.M. Brown, "Automatic Test Performance Criteria for Item Replacement Under Drift-Fault Conditions." The Automation of Testing, IEE Conference Publication No. 91, pp. 29-34, September 20-22, 1972.

26   J.M. Brown and J.D. Lamb, "Fundamental Properties of the Impulse Response of Low-Order Linear Systems." *Int. Journal of Control*, 9, (2), 1969.

27   F.D. Brown, N.F. McAllister, and R.P. Perry, "An Application of Inverse Probability to Fault Isolation." IRE Trans. Military Elect.

28   L. Buschsbaum, M. Dunning, T.J.B. Hannom, and L. Math, "Investigation of Fault Diagnosis by Computational Methods." Remington Rand, May 1964, AD-601204.

29   L.J. Buchsbaum, "Fault Diagnosis by Computational Methods." MSE Thesis, The Moore School of Electrical Engineering of the Univ. of Pennsylvania, Dec. 1963.

30   D. Cabra, "On-Board Checkout System Concept and Philosophy." NASA Manned Space Center, Automatic Support System for Advanced Maintainability, Nov. 1968.

31   G.A. Cambell, "Collected Papers of George A. Cambell." American Telephone and Telegraph Co. New York, 1937.

32   W.K. Chen and F.N. Chan, "On the Unique Solvability of Linear Active Networks." *Trans. Circuits and Systems*, Vol. CAS-21, January 1974, pp. 26-35.

33   L.G. Chesler and R. Turn, "Some Aspects of Man-Computer Communication in Active Monitoring of Automated Checkout." The Rand Corp., March 1967, AD-648553.

34   L. Chesler and R. Turn, "The Monitoring Task in Automated Checkout of Space Vehicles." The Rand Corporation, Memorandum RM-4678, NASA, Sep. 1965.

35   J.P. Chorzel, J.R. Thompson, and R.G. Myers, "System Parameter Measurement Using Transient Response Analysis." Dynamic Testing, A Technique for Automatic Checkout of Closed-Loop Systems, Contract NASW410, General Electric Apollo Systems Department, April 1968.

36   A. Combet, "Application du Programme Sideral au Calcul des Defaillances Par Derive d'un Equipement Analogique Lineaire." Annales de Telecommunications, Vol. 27, No. 3, pp. 111-22, March 1972.

37   A. Combet, "Les Programmes Arcnet et Sideral D'analyse des Reseaux et de Calcul Previsionnel des Defaillances Par Derive D'Equipements Analogiques Lineaires." *Journees D'etude sur les Donnees de Fiabilite, Onet Lannion*, June 10-11, 1971, p. 266.

38   J. Connell, "Analog Circuits:Designing an Adequate Test." IEEE Intercon, 1974.

39   J.L. Constanza and R.L. Osborne, "Fault Detection and Diagnosis, An Energy Point of View." Proc. of Aut. Support Syst. Symp. for Adv. Maintainability, June 1965.

40   G.C. Curtis and J.T. Mash, "An Airborne Manual/Automatic Malfunction Detection System." Lockheed Georgia Company, Marietta, Georgia.

41   J. Dent, " Diagnostic Engineering." *IEEE Spectrum*, Vol. 4, July 1967, p. 99.

42   R. Dobriner, "ACE: The Ultimate in Failure Detection." *Electronic Design*, Vol. 24, November 22, 1967.

43   S.M. Drezner and O.T. Gatt, "Computer-Assisted Countdown." The Rand Corporation for NASA, Memorandum RM-4565, NASA, May. 1965.

44   G. Dube and J.C. Rault, "La Detection et la Localisation des Defauts dans les Circuits Analogiques." *Revue Technique Thomson*, CSF, Vol. 6, March 1974, pp. 71-80.

45   J.P. Dyer, "Identification of Faulty Components of Linear Networks." Naval Postgraduate School, Star, 01 P0066, N70-10706, 1970.

46   Dynamic Testing, "A Technique for Automatic Checkout of Closed-Loop Systems." Volume I and II, Apollo Systems Department, General Electric, Hunstville, Alabama.

47   M. Eleccion, " Automatic Testing: Quality Raiser, Dollar Saver." *IEEE Spectrum*, pp. 38-43, August 1974.

48   S. Even and A. Lempel, "On a Problem of Diagnosis." *IEEE Transactions on Circuit Theory*, Vol. CT-14, No. 3, pp. 361-4, September 1967.

49   G. S. Fang and T. Pavlidis, "Signal Classification Through Quasi-Singular Detection with Applications in Mechanical Fault Diagnosis." *IEEE Transactions on Information Theory*, Vol. IT-18, No. 5, pp. 631-36, September 1972.

50   "Fault Isolation by Parameter Identification." Final Report, RCA Corporation, Defense Electronics Products, Aerospace Systems Division, Burlington, Mass. CR-70-588-68, 1970.

51   W. G. Faust, "The Application and Development of Search Techniques for Fault
      Isolation." Ph.D. Dissertation, University of Pennsylvania, May 1969.

52   S. I. Finkel, R. N. Nilson, and E. S. T. Clair, "A Mathematical Automatic Fault
      Isolation in a Complex System." Vitro Laboratories, AD-413 305, May 1963.

53   Sidney I. Firstman and B. Gluss, "Optimum Search Routines for Automatic Fault
      Location." 15th Meeting ORSA, Washington, D.C., May 1959.

54   Sidney I. Firstman and B. Gluss, "Current Concepts and Issues of In-Space
      Support." Rand Corporation, Santa Monica, Calif., August 1964, AD-606580.

55   S.M. Fisch and G.R. Brigida, "Computer Algorithm for Fault Isolation and Test
      Point Selection." RCA, Burlington, Mass., December 1964, AD-613960.

56   T. Fukui and Y. Nakanishi, "A Method of Designing Output Sensors for
      Equipment Diagnosis Based on Information Content of Output Sensors."
      Electronics and Communications in Japan, Vol. 51-3, Jan. 1968, pp. 113-121.

57   U.R. Furst, "Automatic Built-In Test of Advanced Avionics Systems-Part
      II-Design Techniques for Automatic Built-In Test." *TAES*, July 1966.

58   M.R. Garey, "Optimal Binary Decision Trees for Diagnostic Identification
      Problems." Ph.D. Dissertation, University of Wisconsin, 1970.

59   R.F. Garzia, "Sensitivity Analysis of Fault Isolation Computer Methods." Part I
      Technical Report, Computer Sciences Corporation, NASA Contract NAS8-18405,
      May 1971.

60   R.F. Garzia, "Fault Isolation Computer Methods." NASA Contractor Report
      CR-1758, Computer Sciences Corp., Marshall Space Flight Center, Huntsville,
      Alabama, July 1970.

61   R.F. Garzia, "A Fast Technique for Dynamic Fault Detection." 1970 IEEE
      Automatic Support System Conference pp. 202-206.

62   R.F. Garzia, "Fault Isolation Computer Methods." NASA Contract Report
      CR-1758, Computer Sciences Corporation, February 1971.

63   R.F. Garzia, "Dynamic Methods for Automatic Fault Detection in Continuous
      Linear Systems." IEEE Eurocon 1971, Lausanne, Switzerland.

64   R.F. Garzia, "Fault Isolation in Complex Systems via Bode Diagram Tech-
      niques." Automatic Support Systems Symposium for Advanced Maintainability,
      1972.

65   R.F. Garzia, "System Identification Using a Modified Laplace Transform." 7th
      International Congress on Cybernetics, Namur, Belguim, 10-15, September 1973.

66   R.F. Garzia, "Mathematical Technique for On-Line Determination of Transfer
      Function Coefficients." Seventh Asilomar Conference on Circuits, Systems, and
      Computers, Pacific Grove, California, November 21-29, 1973.

67   R. F. Garzia, "On the Diagnosis of Controllable and Observable Systems." 1971
      International IEEE Conference on Systems, Networks and Computers, January
      19-21, 1971, Oaxtepec, Morelos, Mexico.

68   R. F. Garzia, "Comparison Between Fast Techniques for Dynamic Fault Detection
      in Linear Systems." The Third Annual Southeastern Symposium on System
      Theory, The Georgia Institute of Technology, April 5-6, 1971, Atlanta, Georgia.

69   R. F. Garzia, "COMAD-Application to Future Manned Space Vehicles." 39th
      National Meeting of ORSA, Dallas, Texas, May 5-7, 1971.

70   R. L. Gayer, "Fault Isolation in Solid State Circuits." MSE Thesis, The Moore
      School of Electrical Engineering of the University of Pennsylvania, August 1963.

71 General Electric, "Dynamic Testing-A Technique for Automatic Checkout of Closed-Loop Systems." Contract NASW-410, General Electric, Apollo Systems Department, April 1968.

72 B. Gluss, "An Optimum Policy for Detecting a Fault in a Complex System." 15th Meeting ORSA, Washington, D. C., May 1959.

73 T. P. Goodman, and J. B. Reswick, "Determination of System Characteristics from Normal Operation Records." Trans. ASME 78, February 1956.

74 D. M. Goodman, "Nondestructive Testing of Electronic Circuits with Fiber Optic Scientillators Vidicon Data Sampling and Pat. Reg.", 1966.

75 D. M. Goodman, "A Review of the State of the Art in Automatic Electronic Test Equipment." Reprint Paper No. 389, International Convention, 1966.

76 A. M. Greenspan and L. J. Rytter, "Computer Aided Test Generation for Analog Circuits." 1973 Wescon, pp. 10/4-1 to 10/4-7.

77 S. E. Grossman, "Automatic Testing Pays Off . . ." , *Electronics*, September 19, 1974, pp. 95-109.

78 W. J. Hankley and H. M. Merrill, "A Pattern Recognition Technique for System Error Analysis." *IEEE Transactions on Reliability*, Vol. R-20,No. 3,pp. 148-153, August 1971.

79 W. W. Happ and F. J. McIntosh, "Program Seal-Subnetwork Enumeration and Listing." NASA Techbrief, 68-10227.

80 W. W. Happ, "Combinational Analysis of Multiterminal Devise." *IEEE Transactions on Systems Science and Cybernetics*, Vol. SSC-3, No. 1, pp. 21-27, June 1967.

81 W. W. Happ, "Identification of Test Points in Devices with Specified Symmetry." Proceedings of the International Symposium on CAD, Southampton, April 1969, pp. 496-505.

82 W. W. Happ and E. Sakrisian, "Combinatorial Techniques for Fault Identification in Multiterminal Networks." Proceedings of the Annual Symposium on Reliability 1968, *IEEE Transactions on Reliability*, Vol. , No. 1, pp. 447-85, January 1968.

83 G. E. Harland, M. A. Panhko, K. F. Gill, and J. Schwarzenback, "Pseudo Random Signal Testing Applied to a Diesel Engine." Vol. 13, No. 128, pp. 137-140, 1969.

84 D. Hausrath and R. Ranalli, "Computer Studies of Abnormally Operating Circuits." Proceedings of the 1966 Annual Symposium on Reliability and Quality Control.

85 J. K. Hay and J. M. Blew, "Dynamic Testing and Computer Analysis of Automotive Frames." Eng. Congress, Detroit, January 1972.

86 S. Hayashi, Y. Hattori and T. Sasaki, "Considerations on Network Element Value Evaluation." *Electronics and Communications in Japan*, Vol. 50, Dec. 1967, (transl. Oct. 1968), pp. 118-127.

87 B. Hristoua, A. Angelov, and J. Marinov, "On Some Possibilities of Applying Adaptive Procedures of Failures in Electronic Circuits." 7th International Congress on Cybernetics, Namur, Belgium, September 10-15, 1973.

88 P. H. Jackson, "Fault Isolation by Transfer Characteristics." MSE Thesis, The Moore School of Electrical Engineering of the Univ. of Pennsylvania, May 1969.

89 R. L. Jaegly, "Test Procedure Validation by Computer Simulation." AIAA Second Flight Test Simulation and Support Conference, Los Angeles, California, 1968.

90   J. L. Johnson, H. R. McKenzie, and D. H. Moore, "On-Board in Flight Checkout Evaluation." IBM Federal Systems Division, June 1968, AD-883685.

91   R. A. Johnson, E. Kletsky, and J. Brule, "Diagnosis of Equipment Failures." Syracuse University Research Institute, April 1959.

92   J. L. Johnson, "On-Board in Flight Checkout Part I." IBM Federal Systems Division, December 1966, AD-807534.

93   W. G. Kahn, "Computer Diagnostics and the Computer Chassis Analyzer (MARCOMP)." IR 307, Martin Company, Orlando, Florida, August 1962.

94   T. Killin and R. E. Tulloss, "Automatic Test Systems." IEEE Spectrum, 1974.

95   H. D. Kimp, "Computer Aided Test Design Test." 1973 WESCON, pp. 10/1-1 to 10/1-8.

96   J. E. D. Kirby, D. R. Towill, and K. J. Baker, "Transfer Function Measurement Using Analog Modeling Techniques." *IEEE Trans. on Instrumentation and Measurement,* Vol. IM-22, No. 1, pp. 52-61, March 1973.

97   R. Kirkman, "The Relative Effectiveness of Internally Programmed and Sequential Programmed Machines for Automatic Checkout." IEEE.

98   E. J. Kletsky, "Fundamental Limitations on Self-repairing Systems." Ph.D. Dissertation, Syracuse University, 1961.

99   T. J. Kobylarz, F. H. Herman, A. J. Graf, and P. C. Calella, "Monitoring IC Process Parameters with Statistically Enhanced Test Pattern Data." Seventh Annual Princeton Conference on Information Sciences and Systems, March 22-23, 1973.

100 P. B. Kraabel, "Power Spectra Analysis as a Means of On-Line Checkout."

101 T. Kranton and A. Libenson, "A Pattern Recognition Approach to Fault Isolation." *IEEE Transactions on Aerospace Support Conference Procedures,* Vol. AS-1, No. 2, pp. 1320-6, August 1963.

102 P. Laclair and J. P. McCarthy, "Transfer Function Testing." Automatic Support Systems Symposium for Advanced Maintainability, November 13-15, 1972.

103 J. D. Lamb, "Dynamic Plant Adjustment Using Time Domain Sensitivity Criteria and Cross-Correlation Techniques." IFAC Symposium on Digital Simulation of Continuous Processes, Gyor, Hungary, September 1971.

104 V. S. Levaldi and L. D. Turner, "Fault Diagnosis by White." IEEE Conference on Military Electronics, 1965.

105 V. S. Levaldi, "Pattern Recognition Applied to Fault Detection Noise Techniques." Honeywell, Inc., Minneapolis, Minnesota, May 1966.

106 H. Levenstein, "Use Difference-Equations to Calculate Frequency Response from Transient Response Control Engineering." April 1957.

107 E. Levinson, "A Direct Method of Fault Isolation." MSE Thesis. The Moore School of Electrical Engineering of the University of Pennsylvania, May 1969.

108 F. Liguori, Ed., "Automatic Test Equipment: Hardware, Software and Management." IEEE Press, New York, October 1974.

109 M. F. S. Lin and S. P. Chan, "Fault Diagnosis of Linear Systems." Proceedings of the 7th Annual Allerton Conference on Circuit and System Theory, 1969, pp. 503-10.

110 K. Loewenstein, "Design of Automatic Functional Tester for a Miniaturized AM Radio Integrated Circuit." MSE Thesis, The Moore School of Electrical Engineering, University of Pennsylvania, August 1969.

111   R. Lohse, R. and L. Lauler, "Vade-A System for Real-Time Space Vehicle Checkout and Launch Monitoring." *IEEE Transactions ASCP*, 1969.

112   F. Lu, "Ph.D. Dissertation, National Taiwan University." Taipet, 1972.

113   J. Lustig and D.M. Goodman, "Trends in the Development of Automatic Test Equipment." Project SETE, Report 210/106, National Aeronautics and Space Administration, June 1973.

114   P.A. Lux, "An Adaptive Machine for Fault Detection of Process Control." *IEEE Transactions on Industrial Electronics and Control Instrument*, December, 1967.

115   A.E. Mace, R.N. Pesat, R.T. Minkoff, J.B. Wertz, and A.H. Lipis, "Formulation of System Status Control Techniques." September, 1963.

116   J.H. Maenpaa, C.J. Stehman and W.J. Stahl, "Fault Isolation in Conventional Linear Systems." A Progress Report, *IEEE Transactions on Reliability*, Vol. R-18, No. , p. 12-14, February 1969.

117   L. Mah, "Fault Isolation of Electronic Circuits Using Transfer Function Methods." AD-608 167, October 1964.

118   M.L. Mah, L. Buchsbaum and T.J. Hannom, "Investigation of Fault Diagnosis by Combinational Methods for Microcircuits." AD-623 957, November 1965.

119   J.J. Marraway, "Accessibility in Systems." The Automation of Testing, IEE Conference No. 91, September 20-22, 1972, pp. 1-6.

120   G. O. Martens and J. D. Dyck, "Fault Identification in Electronic Circuits with the Aid of Bilinear Transformations." *IEEE Trans. on Reliability*, Vol. R-21, No. 2 pp. 99-104, May, 1972.

121   Martin Marietta, "Proposal for Investigation of Mathematical Techniques for Fault Isolation." Martin Marietta, ER12206P, January 1962.

122   L. Massa, "A Probabilistic Procedure for the Identification of Errors in Continuous Data." 44th National ORSA Meeting, November 12-14, 1973.

123   L. T. Mast, "Prelaunch Checkout in the 1970's." The Rand Corporation, Santa Monica, California, April 1965, AD-614410.

124   L. T. Mast, "Growth of Automation in Prelaunch Checkout for Space Vehicles." The Rand Corporation, Santa Monica, California, May 1967, AD-652458.

125   L. Math, L. Buchsbaum, and T. J. B. Hannon, "Investigation of Fault Diagnosis by Computer Methods for Microcircuits UNIVAC." Blue Bell, Pennsylvania, November 1965, AD-623951.

126   W. R. McCormack, "Analog Checkout of Large Systems—The Digital Solution." NEC, October 1962.

127   W. R. McCormack and C. Michel, "Diagnostic Maintenance, A Technique Using a Computer." *IEEE Transactions on Aerospace*, Vol. AS-1, pp. 931-41, August 1963.

128   F. J. McIntosh and W. W. Happ, "Programmed Test Patterns for Multiterminal Devices." *SJCC*, 1969, pp. 229-40.

129   McSweeney, K. F., "Malfunction Detection System for Advanced Spacecraft." *IEEE Transaction on Aerospace and Electronics Systems*, January 1966.

130   H. M. Merrill, "N10 Failure Diagnosis Using Pattern Recognition." M.S. Thesis, Department of Electrical Engineering, University of Utah, August 1968.

131   H.M. Merrill, "Failure Diagnosis Using Quadratic Programming." *IEEE Transactions on Reliability*, Vol. R-22, No. 4, pp. 207-213, October 1973.

132   W.D. Moon, "Periodic Checkout and Associated Errors." IEEE Transactions on
      Aerospace, Vol. 2, RCA, Burlington, Mass., April 1964.

133   I.E. Morse, W.R. Shapton, D.L. Brown, and E. Kuljanic, "Applications of Pulse
      Testing for Determining Dynamic Characteristic of Machine Tools." 13th Inter.
      Machine Tool Design and Research Conference, University of Birmingham,
      England, September 1972.

134   D. Nalley, "Preliminary Study of Some Fault Isolation Techniques." Computation
      Lab Sci. Eng. Directorate, Marshall Space Flight Center, Huntsville, Alabama.

135   A.C. Nelson, C.A. Krohn, W.S. Thompson, and J.R. Batts, "Evaluation of
      Computer Programs for System Performance Effectiveness." Research Triangle
      Institute, North Carolina, Report TRI-SU265, 1966.

136   E.C. Neu, "A New n-Port Network Theorem." Proc. of the 13th Midwest Symp. on
      Circuit Theory, Minneapolis, 1970 (paper iv. 5).

137   E.C. Neu, "Combinatorial Analysis for Checkout Techniques." Joint National
      Meeting of the American Astronautical Society.

138   G.V. Novatny, "Automatic Checkout Equipment." Electronics, pp. 37-44, July 13,
      1962.

139   D.E. Olson and S.G. Boyer, "Failure Diagnosis Program of N10 Guidance
      System." System Research Laboratory, Department of Electrical Engineering,
      University of Utah, February 1966.

140   R.L. Osborne, "The Detection and Diagnosis of Malfunction in Energy
      Manipulating Systems." University of California, Berkeley, December 1967.

141   T. Pavlidis and G.S. Fang, "Application of Pattern Recognition to Fault Diagnosis
      of Internal Combustion Engines." IEEE Computer Group Repository, R-72-20.

142   T. Pavlidis and G.S. Fang, "A Segmentation Technique for Waveform
      Classification." IEEE Transactions on Computers, Vol. C-21, No. 8, pp. 901-904,
      August 1972.

143   P.A. Payne, D.R. Towill and K.J. Baker, "Predicting Servomechanism Dynamic
      Response from Limited Production Test Data." Radio and Electronic Engineer,
      Vol. 40, No. 6, December 1970, pp. 275-88.

144   Programming, "Integration and Checkout of the Advanced Radar Traffic Control
      System." Univac Division of Sperry Rand Corporation, April 1965, AD-622865.

145   N.N. Puri and C.N. Weygandt, "Transfer Function Tracking of Linear Time
      Varying by Means of Auxiliary Simple Lag Networks." Joint Acc, 1963.

146   R. Ranalli, "Automatic Fault Isolation Techniques for Nondigital Applications."
      Proceedings of the 1968 IEEE International Convention Record, p. 196.

147   M.N. Ransom and R. Saeks, "Fault Isolation with Insufficient Measurements."
      IEEE Transactions on Circuit Theory, Vol. CT-20, No. 4, pp. 416-417, July 1973.

148   M.N. Ransom, "A Functional Approach to the Connections of a Large Scale
      Dynamical System." Ph.D. Dissertation, Notre Dame University, May 1973.

149   M.N. Ransom, and R. Saeks, "Fault Isolation via Term Expansion." 3rd Pittsburgh
      Symposium on Modeling and Simulation, Univ. of Pittsburgh, Vol. 3, pp. 224-228,
      1973.

150   J.C. Rault, "Depannage Automatique des Circuits Electroniques Analogiques."
      Thomson, CSF, LCR/DRS, Note Technique No. 1521, 22 Avril, 1971.

151   E. Rivera, E.R. Garcia, and R. Ranalli, "Computer Generated Fault-Isolation

Procedures." Proceedings of the 1967 Annual Symposium on Reliability and Quality Control.

152   J.E. Rodriguez, "Circuit Simulation as a Fault Analysis Tool, IEEE International Symposium on Circuit Theory, April 9-11, 1973.

153   B. Rozenwaig, "Un Programme de Calcul Previsionnel des Defaillances Par Derive des Circuits Analogiques Lineaires." Colloque International sur la Microelectronique Avancee, 6-10 Avril 1970, Tome 2, Editions Chiron 1970, p. 1261.

154   R. Saeks, "Fault Isolation, Component Decoupling and the Connection Groupoid." Computation Laboratory, MSFC, August 21, 1970.

155   R. Saeks, S.P. Singh, and R.W. Liu, "Fault Isolation via Components Simulation." *IEEE Transactions on Circuit Theory*, Vol. CT-19, No. 6, pp. 634-640, November, 1972.

156   E. Sakrisian and W.W. Happ, "Combinatorial Techniques for Fault Identification in Multiterminal Devices." Proceedings of the Annual Symposium on Reliability, pp. 477-85, January 1968.

157   S.P. Schechnea, "Management Strategy and Policy in Fault Isolation Techniques." MBA Thesis, Wharton School of the University of Pennsylvania, 1968.

158   A. Sen and J.C. Strauss, "Parameter Determination from Experimental Tests." Proceedings of the Midwest Symposium of Circuit Theory, 1969, p.

159   S. Seshu and R. Waxman, "Fault Isolation in Conventional Linear Systems-A Feasibility Study." *IEEE Transactions on Reliability*, Vol. R-15, No. I, pp. 11-16, May 1966.

160   G. Shapiro, G.J. Rogers, O.B. Lang and P.M. Fulcomer, "Project Fist Fault Isolation by Semiautomatic Techniques." Part-1, Basic Concept and Techniques, Part-2, Detailed Instrumentation, *IEEE Spectrum*, Vol. I, pp. 98-111 and pp. 130-144, August and September 1964.

161   S.P. Singh, "Ph.D. Dissertation." University of Notre-Dame, Indiana, 1971.

162   E.S. Sogomonyan, "Monitoring Operability and Finding Failures in Functionally Connected Systems." Translated from Avtomatika i. Telemekhanika, Vol. 25, No. 6, June 1964, pp. 980-990 (Moscow).

163   S. Srivastava, "Development of Fault Isolation Procedures for Semi-automatic Fault Isolation of Electronic Assemblies." MSE Thesis, The Moore School of Electrical Engineering of the Univ. of Pennsylvania, Aug. 1968.

164   H. Sriyananda, "Application of the Method of Potential Functions to Fault Diagnosis." *Electronics Letters*, Vol. 8, No. 6, March 1972, pp. 159-60.

165   H. Sriyananda and D. R. Towill, "Fault Diagnosis via Automatic Dynamic Testing—A Voting Technique." The Automation of Testing, IEE Conference Publication No. 91, pp. 196-201, September 20-22, 1972.

166   W. J. Stahl, T. K. McBride, and J. H. Maenpaa, "Investigation of Fault Isolation Diagnosis by Transfer Function Techniques." ITT Kellogg Communications Systems, Chicago, Illinois, Report on Contract No. AF 33(615) 2302, November 1965.

167   W. T. Stahl, "Investigation of Fault Diagnostic by Computer Methods for Microcircuits." AFAPL, TR-65TTT, AD-625 587, November 1965.

168   W. J. Stahl, J. H. Maenpaa, and C. J. Stehman, "Development of Advanced Dynamic Fault Diagnosis Techniques." Technical Report AFAPL, TR. 67-44, May

1967, Report on Contract No. AF33(615)-3573, Scully International Inc., AD-257 887.

169   W. J. Stahl, J. H. Maenpaa, and C. J. Stehman, "Computer-Aided Design Part 13—Defining Fault with a Dictionary." *Electronics*, 22 January, 1968, pp. 64-68.

170   W. J. Stahl and J. H. Maenfaa, "Development of Advanced Dynamic Fault Diagnosis Techniques." Scully International, Inc., May 1967, AD-814457.

171   W. J. Stahl, "Dynamic Fault Diagnosis Techniques." Scully International Inc., Downers Grove, Illinois, September 1968.

172   L. St. Clair, S. I. Finkel and R. N. Nelson, "A Mathematical Technique for Automatic Fault Isolation in Complex Systems." Vitro Labs, W. Orange, N. J., Final Report on Contract AF-33 (657)-9184, May 1963.

173   C. J. Stehman, J. H. Maenpaa, and W. J. Stahl, "Complete Tree Generation—Some Practical Applications." *IEEE Transactions on Circuit Theory*, Vol. CT, No., pp. 548-550, November 1969.

174   A. Stewart, "Statistical Modelling for Field Drift Testing." The Automation of Testing, IEE Conference Publication No. 91, September 20-22, 1972, pp. 87-92.

175   A. D. Swain and J. G. Whol, "Factors Affecting Degree of Automation in Test and Checkout Equipment." The Rand Corporation, March 1961, AD-257887.

176   J. W. Taylor, "Automatic Checkout Systems for Combat Vehicles." Frankford Arsenal, Philadelphia, Pennsylvania, April 1964, AD-823067.

177   A. R. Teasdale and J. B. Reynolds, "Two Ways to Get Frequency Response from Transient Data." *Control Engineering*, October 1955.

178   E. F. Thomas, "DC Pin-to-Pin Testing of Integrated Circuits." NASA/GSFC Paper 515-001, May 1967.

179   K. To and R. E. Tulloss, "Automatic Test Systems." *IEEE Spectrum*, Sept. 1974, pp. 44-52.

180   D. R. Towill and P. A. Payne, "Frequency Domain Approach to Automatic Testing of Control Systems." *The Radio and Electronic Engineer*, Vol. 41, No. 2, Feb. 1971.

181   J. R. Townes, S. J. Dwyer, R. W. McLaren, and G. W. Zobrist, "Linear Techniques for Binary Fault Isolation." University of Missouri.

182   UNIVAC, "Investigation of Fault Diagnosis by Computational Methods." Final Report APL TDR 64-62 on Contract AF 33(657)-10113, May 1964.

183   J. E. Valstar, "In Flight Dynamic Checkout." *IEEE Transactions on Aerospace Support Conference*, August 1963.

184   J. E. Valstar, "Fault Isolation of Electronic Circuits Using Transfer Function Methods." AF Aero Propulsion Lab, Wright Patterson, Ab, Ohio, October 1964, AD-608 176.

185   J. E. Valstar, "Some Fundamental Limitations on Indirect Testing of Dynamic Systems." Supp. *IEEE Transactions on Aerospace and Electronic Systems.* Vol. AES-2, No. 4, pp. 455-62, July 1966.

186   M. M. Vartanian, "Validation Study of an Electronic Network Fault Isolation Procedure." MSE Thesis, The Moore School of Electrical Engineering of the University of Pennsylvania, December 1967.

187   M. M. Vartanian, "An Algorithm for Fault Isolation of Multi State Electronic Networks." Ph.D. Dissertation, University of Pennsylvania, 1969.

188 A. S. Weitzenfeld and W. W. Happ, "Combinational Techniques for Actions on Reliability." Vol. R16, No. 3, pp. 93-99, December 1967.

189 J. L. Willows and W. G. Magnuson, "Bias—A Network Analysis Computer Program Useful to the Reliability Engineer." *IEEE Transactions on Reliability*, Vol. R-20, No. 3, pp. 108-110, August 1971.

190 J. K. Wolfe and J. H. Dietz, "A Statistical Theory for Parameter Identification in Physical Systems." *Journal of the Franklin Institute*, pp. 369-400, November 1962.

191 P. J. Wong, "Application of Decision Theory to the Testing of Large Systems." *IEEE Transactions on Aerospace and Electronic Systems*, Vol. AES-7, No. 2, pp. 378-384, March 1971.

192 S. M. Worthington, "Application of Automatic Checkout and Fault Isolation Techniques to Automatic Control Systems." MSE Thesis, The Moore School of Electrical Engineering of the University of Pennsylvania, December 1967.

193 K. H. Yim, "Fault Identification Matrix In Linear Networks." AD-721 581, September 1970.

A BIBLIOGRAPHY ON FAULT DETECTION
AND LOCATION IN DIGITAL SYSTEMS

Compiled by
Jean-Claude Rault

Test Et Diagnostic

1 S. K. Abdali, "On Proving Sequential Machine Design," *IEEE Transactions On Computer*, Vol. C-20, No. 12, p. 1563-6, December 1971.

2 H. G. Adshead, G. C. Jain and A. J. Knowles, "New Dimensions in Automatic Logic Testing and Diagnostics," International Conference on Computer-Aided Design, Southampton, England, 24-28, April 1972, p. 112-18.

3 V. D. Agrawal and P. Agrawal, "An Automatic Test Generation System For Illiac IV Logic Boards," *IEEE Transactions on Computers*, Vol. C-21, No. 9, p. 1015-1017, September 1972.

4 B. Aguilhon and J. P. D'Issernio, *"Test De Systemes Sequentiels Par Sequences Binaires Pseudo-Aleatoires,"* Projet D'Eleves Laboratoire D'Automatique De Grenoble, 1970.

5 S. B. Akers, "On a Theory of Boolean Functions", *Journal of the Society of Industrial Mathematics*, Vol. No. 4, p. 487-98.

6 S. B. Akers, "A Rectangular Logic Array", *IEEE Transactions On Computers*, Vol. C-21, No. 8, p. 848-57, August 1972.

7 S. B. Akers, "Universal Test Sets For Logic Networks", Proceedings of the IEEE 1972 Symposium on Switching and Automata Theory, pp. 177-184.

8 S. B. Akers, "Universal Test Sets For Logic Networks," *IEEE Transactions on Computers*, Vol. C-22, No. 9, pp. 835-839, September 1973.

9 S. Akers, Diagnosis and Graph Coloring Workshop on Diagnosis and Reliable Design of Digital Systems, Pasadena, California, December 5-7, 1973.

10 G. P. Aksenova and E. S. Sogomonyan, "Synthesis of Built-In Test Circuits for Automata With Memory", *Automation and Remote Control*, Vol. 32, No. 9, p. 1492-1500, September 1971.

11 O. G. Alekseev and V. G. Staroselet, "Setting-Up Algorithms Intended For the Optimal Selection of Parameters of Complex Systems When Checking Their Operability", *Automation And Remote Control*, Vol., No. 9, p. 1631-1638, 1965.

12 O. G. Alekseev, *et al.*, "The Optimal Selection of Parameters for Testing Systems with Complex Servicing Sequences," *Automation and Remote Control*, No. 11, p. 107-114, 1966.

3 S. Alexander, "Application of Boolean Notation to the Maintenance of Switching Circuits", *Electrical Engineering*, Vol. 33, No. 6, p. 372-374, June 1961.

4 J. Allain, J. Y. Meuric, J. F. Perrin and P. Pignal, "Une Methode Automatique De Depannage Des Ensembles Logiques", *L'Echo Des Recherches*, No. 57, p. 50-7, July 1969.

15 N. C. Allegre, "Tesla, A Control Language For Logic Simulations Of Digital Circuits", Illiac IV Document No. 187, Department of Computer Science Report No. 347, Urbana, Illinois, University of Illinois, August 1969.

16 D. P. Allen and N. P. Lyons, "Selecting Inputs To Test Digital Circuits On Complex IC Boards," *Electronics*, July 17, 1972, p. 88-92.

17 D. Alonzo, "Diagnostic Testing In Production", Proceedings of the Symposium-Testing to Integrate Semiconductor Memories into Computer Mainframes, October 4, 1972.

18 O. Altan and B. Borgerson, "Automatic Test Generation Utilizing A Mixed Resolution System Simulator," Workshop on Diagnosis and Reliable Design of Digital Systems, Pasadena, California, December 5-7, 1973.

19 V. Amar and N. Condulmari, "Diagnosis of Large Combinational Networks", *IEEE Transactions on Electronic Computers*, Vol. EC-16, No. 5, p. 675-80, October 1967.

20 N. G. Anantha, C. S. Chang and J. A. Naselow, "Testing Defective Serial Shift Register", IBM Technical Disclosure Bulletin, Vol. 15, pp. 1961-1962, November 1972.

21 K. Anderson, "Test Problems In Device Manufacturing", Proceedings of the Symposium—Testing to Integrate Semiconductor Memories into Computer Mainframes, October 4, 1972.

22 R. M. Apte, "Derivation Of Near Minimal Test Sets And Fault Partitioning Techniques", Ph.D. Thesis, Southern Methodist University, Dallas, Texas, August 1973.

23 I. Arakami, "Design Checkout Of Logic Networks", IEEE Computer Group Society Repository R-71-14.

24 D. B. Armstrong, "On Finding A Nearly Minimal Set Of Fault Detection Tests For Combinational Logic Nets," *IEEE Transactions on Electronic Computers*, Vol. EC-15, No. 1, p. 66-73, February 1966.

25 D. B. Armstrong, "A Logic Fault Simulator For The Sentinel Data Processing System", Bell Telephone Lab Internal Memo, 1968.

26 D. B. Armstrong, "A Deductive Method For Simulating Faults In Logic Circuits", *IEEE Transactions On Computers*, Vol. C-21, No. 5, p. 464-71, May 1972.

27 F. T. Arnold, "Automatic Testing of Wiring For Circuit Card Panels", J. Acc., p. 49-53, 1963.

28 A. H. Ashkinazy, "Fault Detection In Asynchronous Sequential Machines", Doctoral Thesis, Columbia University, New York, July 1970.

29 A. Ashkinazy, "Fault Detection Experiments In Asynchronous Sequential Machines", Proceedings of the Eleventh Annual Symposium On Switching and Automata Theory, 1970, p. 88-96.

30 I. O. Atovmyan, M. I. Arshavskii, Y.G. Drevs, and V. V. Zolotarev, "Some Problems In The Development Of Programs For The Diagnosis Of Malfunctions In Devices In A Digital Computer System," AD 762237, May 31, 1973, Edited Translation of *Vsesoyuznoe Soveshchanie Po Tekhnicheskoi Diagnostike Trudy*, 1972, pp. 57-61.

31 P. Azema, *"Difference Booleenne D'Une Fonction Logique Monotone, Cr Academie Des Sciences"*, Tome 268 Serie A, p. 473-4, February 24, 1969.

32 P. Azema, M. Courvoisier, and J. P. Richard, "Fault Detection Method In Sequential Systems", N72-29168, 1971, also Laas Publication No. 912.

33 P. Azema, M. Courvoisier, and J. P. Richard, *"Methode De Detection Des Pannes Dans Les Systemes Sequentiels,* 5 Eme Congres Mondial IFAC, Paris, June 1972, p. 167.

34 J. D. Baker, "The Fault-Cover Problem In Combinational Logic Circuits", M. Sc. Thesis, Department of Electronics, University of Southampton, 1971.

35 M. Ball and F. Hardie, "Effects And Detection Of Intermittent Failures In Digital Systems", FJCC 1969, p. 329-335.

36 M. R. Baray, "An Algorithm For Fault Diagnosis In Combinatorial Logic, International Computing Symposium," Proceedings, Cini Foundation, Venice, April 12-14, 1972, p. 593-603.

37 M. Barber, "Techniques For Dynamically Testing Semiconductor Memory Chips," Proceedings of the Symposium. Testing to Integrate Semiconductor Memories Into Computer Mainframes, IEEE Computer Society Mideastern Area Committee, October 4, 1972.

38 R. F. Barry, R. S. Fisher, and R. R. Mattison, "Programmed Algorithm For Test-Point Selection And Fault Isolation", Vol. I Et II, RCA Internal Report, Burlington, Massachusetts, August 1966.

39 T. R. Bashkow, "Review Of The Proceedings Of The Seminar On Automatic Checkout Techniques", September 5-7, 1962, Battelle Memorial Institute, Columbus, Ohio.

40 T. R. Bashkow, J. Friets, and A. Karson, "A Programming System For Detection And Diagnosis Of Machine Malfunctions", *IEEE Transactions*, Vol. EC-12, No. 1, pp. 10-17, February 1963.

41 J. C. Bassett and C. R. Kime, "Improved Procedures For Determining Diagnostic Resolution", *IEEE Transactions on Computers*, Col. C-21, No. 4, p. 385-388, April 1972.

42 D. A. Bastin, E. M. Girard, and J. C. Rault, "A Set Of Interactive APL Programs For Fault Detection And Location In Digital Circuits", 1972 International Symposium on Fault-Tolerant Computing, 19-21 June 1972, p. 47-52.

43 D. Bastin, E. Girard, and J. C. Rault, "Verification Des Sequences De Test Et Determination De Leur Efficacite", Colloque International Conception Et Maintenance Des Automatismes Logiques Toulouse, September 27-28, 1972, Also N73-12198.

44 D. Bastin, E. Girard, J. C. Rault, and R. Tulloue, "Probabilistic Test Generation Methods—Statistical Estimation Of Sequence Length And Efficiency", Note Technique CCTI-SAS No. 1820, Decembre 1972.

45 R. P. Batni, J. D. Russell, and C. R. Kine, "Diagnostic Test Set Reduction By Specifying Don't Cares", IEEE Computer Society Repository, R73-186.

46 H. Bauer and G. A. Grotz, "Method Of Fault Location In Logic Networks", IEEE Computer Society Repository, R-72-145.

47 W. H. Beall and H. E. Brown, "Computer-Aided Diagnosis Of Faults In Digital Networks," Nerem 1970, p. 52-53.

48 L. W. Bearnson, "Arithmetic Error Detection In Digital Computers", M. S. Thesis, Department of Electrical Engineering, Syracuse University, May 1965.

49 L. Bearnson and C. C. Carroll, "Design Of Minimum Fault Test Schedules And Testable Realizations For Combinational Logic Circuits", AD-706 231, May 1970.

50 L. W. Bearnson and C. C. Caroll, "On The Design Of Minimum Length Fault Tests For Combinational Circuits", Proceedings of the International Symposium on Fault-Tolerant Computing, 1-3 March 1971, Pasadena, California, p. 1-4.

51 L. W. Bearnson and C. C. Carroll, "On The Design Of Minimum Length Fault Tests For Combinational Circuits", IEEE Transactions on Computers, Vol. C-20, No. 11, p. 1353-6, November 1971.

52 F. Begon and R. Tremolieres, "Elaboration D'Un Modele Logique Du Diagnostic", Note Scientifique No. 5, IRIA, October 1971.

53 B. Beizer, "A New Theory For the Analysis, Synthesis, Cutting And Splicing Of Sequential Switching Networks", AD-842 002, May 1966.

54 J. E. Belt, "A Heuristic Search Approach to Test Sequence Generation for AHPL Described Sequential Circuits," Ph.D. Dissertation, University of Arizona, 1973.

55 Yu K. Belyaev and I. A. Ushakov, "Mathematical Models for Problems of Detecting and Locating Faults", Cybernetics in the Service of Communism, JPRS-30128, p. 159-177, 1964.

56 C. A. Bennett, "Application of a Learning Curve to a Maintenance Problem, 2nd Annual Quality Control Symposium of the Dallas-Fort Section of the ASQ, March 1961.

57 R. G. Bennetts and D. W. Lewin, "Fault Diagnosis of Digital Systems—A Review", Computer, Vol. 4, p. 12-21, 1971.

58 R. G. Bennetts and D. W. Lewin, "Fault Diagnosis of Digital Systems—A Review," Computer Journal, Vol. 13, No. 2, pp. 199-206, May 1971.

59 R. G. Bennetts, "The Diagnosis of Logical Faults," Wireless World, Vol. 17, July 1971, p. 325-8, and August 1971, p. 383-5.

60 R. G. Bennetts, "A Realistic Approach to Detection Test Set Generation for Combinational Logic Circuits", The Computer Journal, Vol. 15, No. 3, p. 238-46, October 1972.

61 R. G. Bennetts, "A Contribution to the Boolean Difference Procedure for Generating Tests for Combinational Logic Circuits", Datafair 1973.

62 L. C. Benning, "Application of a Logic Fault Analyzer to the Manufacture and Maintenance of the Control Data 7600 Computer," 8th Design Automation Workshop, 1971, p. 231-35.

63 I. Berger and Z. Kohavi, "Fault Detection in Fanout-Free Combinational Networks", IEEE Transactions on Computers, Vol. C-22, No. 10, pp. 908-914, October 1973.

64 T. Bessho, "Memory Testing Methods," Electronics and Communications in Japan, Vol. 54, No. 4, April 1971.

65 A. A. Bessonov, "An Optimal Algorithm for the Indication of Failures", Sbornye Trudy Leningradskogo Instituta, 1967, No. 62, p. 27-30.

66 R. Betancourt, "Derivation of Minimum Test Sets for Unate Logical Circuits", AD-720 330, August 1970.

67 R. Betancourt, "Derivation of Minimum Test Sets for Unate Logical Circuits", IEEE Tranactions on Computers, Vol. C-20, No. 11, p. 1264-9, November 1971.

68 G. Bioul and M. Davio, "Taylor Expansions of Boolean Functions and of Their Derivatives", Philips Research Reports, Vol. 27, No. 1, p. 1-6, February 1972.

69 X. Blanchard and X. Lowery, "Maintainability Principles and Practices", McGraw Hill, New York 1969.

70 P. E. Blankenship and P. G. McHugh, "FDP Diagnostic System," AD-741823, February 1972.

71 W. R. Blatchley, "Computer for Core Tests," *Electronics*, September 1, 1969, pp. 90-92.

72 D. H. Blauvelt, "Fault Isolation in a Digital Guidance and Control Computer," AGARD-The Application of Digital Computer to Guidance and Control, November 1970.

73 V. F. Blefary and W. B. Rohn, "Centralized Automatic Trouble Locating and Analysis System", International Conference on Communications, June 11-13, 1973.

74 J. Blythin and K. J. Crook, "Synthesis of Test Patterns for Logic Networks", IEE Conference on Automatic Test Systems, p. 187-199, Birmingham, April 1970.

75 J. Blythin, "A High Level Test Programming Language and the Use of a Logic Simulator for Program Checking", Joint Professional Group E10/IEE/IERE Computer Group Colloquium on Computer Applications to Design, Simulation and Testing of Logic Circuits and Systems, London, 16 November 1971.

76 O. C. Boelens, "Fault Diagnosis in Iterative Arrays," Ph.D. Dissertation, Stevens Institute of Technology.

77 A. M. Bogomolov, *et al.*, "The Theory of Binary Relations Applied to Detecting and Locating Faults in Complex Systems," *Vychislitelnye Metody I Programmirovanie Po Vychislitelnym Mashinami*, p. 103-125 (IZD-Vo Saratovskovo Universiteta, 1966).

78 A. M. Bogomolov and V. A. Tverdokhlovov, "The Conditions for the Existence of Diagnostics Tests", *Kibernetika*, No. 3, p. 9-19, 1968.

79 A. M. Bogomolov and V. Tverdokhlovov, "Conditions for the Existence of Diagnostics Tests for Complex Systems", IFIR Congress 1968, Edinburgh Proceedings, Vol. A. p. 156-159, 1968.

80 T. J. Bohen, "The Modes of Failures of Digital Memories", Workshop on Hardware Faults and the Programmer, Palm Springs, 28-29 June 1968.

81 J. Г. Bohme, "Diagnostic Preset Experiments with Stochastic Automata", *Computing*, Vol. 8, No. 1/2, p. 79-98, 1971.

82 M. Bohus and K. Geher, "Examination of Logic Networks by Computer", *Hiradastechnika*, Vol. 23, No. 7, p. 199-203, July 1972.

83 T. L. Booth, *Sequential Machines and Automata Theory*, J. Wiley, 1967.

84 B. Borgerson and C. V. Ravi, "On Addressing Failures in Memory Systems", International Computing Symposium Proceedings, Cini Foundation, Venice, April 12-14, 1972, p. 40-47.

85 D. C. Bossen, D. L. Ostapko, and A. M. Patel, "Optimum Test Patterns for Parity Networks", FJCC 1970, p. 63-68.

86 D. C. Bossen and S. J. Hong, "Cause-Effect Analysis for Multiple Fault Detection in Combinational Networks", IBM TR-002161, January 25, 1971.

87 D. C. Bossen, D. L. Ostapko, A. Patel, and M. S. Schmookler, "Minimum Test Patterns for Residue Networks," 8th Design Automation Workshop, 1971, p. 278-84.

88 D. C. Bossen and S. J. Hong, "Cause Effect Analysis for Multiple Fault Detection in Combinational Networks", *IEEE Transactions on Computers*, Vol. C-20, No. 11, p. 1252-7, November 1971.

89 D. C. Bossen, D. L. Ostapko, and A, M. Patel, "Test Patterns for Parity Networks", *IBM Technical Disclosure Bulletin*, Vol. 13, No. 10, p. 2897-9, March 1971.

90 D. C. Bossen, D. L. Ostapko, and A. M. Patel, "Optimum Test Patterns for Parity Networks", *Computer Design*, Vol. 11, No. 4, pp. 93-97, April 1972.

91 F. R. Boswell, "Designing Testability into Complex Logic Boards", *Electronics*, August 14, 1972, p. 116-119.

92 F. R. Boswell, "A Diagnostic Computer System for Fault Diagnosis of Digital Circuits", Automatic Support Systems Symposium for Advanced Maintainability, November 13-15, 1972.

93 P. S. Bottorf, R. A. Rasmussen, "A View of User-Oriented Production Test Generation System", 7th Design Automation Workshop, 1970, p. 90-94.

94 P. S. Bottorf, M. E. Schwass, and F. J. Villante, "AV Automatic System Approach to the Problem of Memory Circuit Testing and Fault Diagnosis", Proceedings of the Share-ACM-IEEE Design Automation Workshop, San Francisco, California, p. 95-99, 22-25 June 1970.

95 P. S. Bottorf, "Test Data Generation for the Manufacturing Environment", Lehigh University Workshop on Fault Detection and Diagnosis in Digital Circuits and Systems, 7-9 December 1970, pp. 45-6.

96 P. S. Bottorf, "Test Generation for LSI Circuit Manufacture", Open Workshop on Fault Detection and Diagnosis in Digital Systems, 6-8 December 1971, Lehigh University.

97 W. J. Bouknight, "On the Generation of Diagnostic Test Procedures", Coordinated Science Laboratory, University of Illinois, Urbana, Illinois, Report R-292, May 1966.

98 W. G. Bouricius, W. C. Carter, K. A. Duke, and J. P. Roth, "Interactive Design of Self-Testing Circuitry", Proceedings of the Purdue Symposium on Information Processing, p. 73-80, April 1969.

99 W. G. Bouricius, E. P. Hsieh, G. R. Putzolu, J. P. Roth, R. R. Schneider, and C. J. Tan, "Algorithms for Detection of Faults in Logic Circuits", *IEEE Transactions on Computers*, Vol. C-20, No. 11, p. 1258-64, November 1971.

100 R. Boute and E. J. McCluskey, "Fault Equivalence in Sequential Machines, Symposium on Computers and Automation", Polytechnic Press of the Polytechnic Institute of Brooklyn, 13-15 April 1971.

101 R. Boute and E. J. McCluskey, "Fault Equivalence in Sequential Machines", AD-731, 700, June 1971.

102 R. Boute, "Algorithms for Combinational Fault Equivalence Using LISP", IEEE Computer Society Repository R-72-202, and Technical Note No. 9, Digital Systems Laboratory, Stanford University, Stanford, California, September 1971.

103 R. Boute, "Adaptive Design Methods for Checking Sequences", Technical Report No. 30, Digital Systems Laboratory, Stanford Electronics Laboratories, Stanford University, Stanford, California, July 1972.

104 R. Boute and E. J. McCluskey, "Fault Equivalence in Sequential Machines", In Computers and Automata, p. 483-507, J. Fox Ed., J. Wiley, 1972.

105 R. Boute and E. J. McCluskey, "Fault Equivalence in Sequential Machines", IEEE Computer Society Repository, R-72-203.

106 R. T. Boute, "Distinguishing Sets and Their Use in Checking Sequences", IEEE Computer Society Repository, R73-28.

107 R. T. Boute, "Algebraic Properties of Test Sequences and Fault Relations", IEEE Computer Society Repository, R73-159.

108 R. T. Boute, "Properties of Memory Faults in Sequential Machines", IEEE Computer Society Repository, R73-161.

109 R. T. Boute, "Checking Experiments for Output Faults", IEEE Computer Society Repository, R73-163.

110 R. T. Boute, "Equivalence and Dominance Relations Between Output Faults in Sequential Machines", IEEE Computer Society Repository, R73-164.

111 R. T. Boute, "Adaptive Design Methods for Checking Sequences," IEEE Computer Society Repository, R73-165.

112 R. T. Boute, "Fault Detection in Fundamental-Mode Circuits", IEEE Computer Society Repository, R73-179.

113 R. T. Boute, "Optimal and Near Optimal Checking Experiments for Output Faults in Sequential Machines", IEEE Computer Society, R73-324.

114 E. K. Bowdon, "Digital Network Design Aids, an Integrated Approach", 1970 Share-ACM-IEEE 7th Design Automation Workshop, p. 220-29.

115 A. H. Boyce, "Generating Diagnostic Testing Procedures of Logic Circuits by Digital Computers", Marconi Review, Vol. 33, p. 246-59, 1970.

116 A. H. Boyce, "Computer Generated Diagnosing Procedures for Logic Circuits", IEE Conference on Automatic Test Systems, p. 333-46, Birmingham, April 1970.

117 A. H. Boyce, R. C. Emmerson, D. V. Stringer, and B. G. West, "Simulation of Binary Logic Circuits by Digital Computers", Marconi Review, Vol. 181, p. 121-42, Second Quarter 1971.

118 A. H. Boyce, "A Study of Diagnostic Testing of Sequential Circuits", International Conference on Computer-Aided Design Southampton, England, 24-28 April 1972, p. 309-14.

119 A. H. Boyce, "A Suite of Programs to Aid Logic Testing, The Automation of Testing", IEE Conference Publication No. 91, p. 145-51, September 20-22, 1972.

120 A. J. Boyle, "Testing Mos, Testing Complex Mos—The How and Why", The Electronic Engineer, Vol. 29, No. 10, p. 41-46, October 1970.

121 D. W. Bray and C. P. Hsieh, "ATVG III Computer Program Description", General Electrical Technical Information Series, R69E1S-128, December 1969.

122 D. W. Bray, "Automatic Generation of Test Procedures for Sequential Networks", 1971 International IEEE Computer Society Conference, p. 75-76.

123 D. W. Bray, "The ATVG Program, A Test Vector for Sequential Networks", Open Workshop on Fault Detection and Diagnosis in Digital Systems, p. 42-54, Lehigh University, 6-8 December 1971.

124 D. Breslow, "Automatic Fault Location Using Building-Block Logic", Proceedings of the Sixth National Symposium on Reliability and Quality Control, p. 449-458, 1960.

125 M. A. Breuer, "General Survey of Design Automation of Digital Computer", Proceedings of the IEEE, Vol. 54, No. 12, p. 1708-1721, December 1966.

126 M. A. Breuer, "Hardware Fault Detection", 1968 FJCC, p. 1502-03.

127 M. A. Breuer, "Fault Detection in a Linear Cascade of Identical Machines", 9th Annual Symposium on Switching Theory and Automata Theory, p. 235-43, 15-18 October 1968.

128 M. A. Breuer, "Functional Partitioning and Simulation of Digital Circuits", *IEEE Transactions on Computers*, Vol. C-19, No. 11, p. 1038-1046, November 1970.

129 M. A. Breuer, "Generation of Fault Detection Tests for Sequential Circuits", AD.724-445 and 1970 International Symposium on Fault-Tolerant Computing, p. 18-21, Pasadena, California, 1-3 March 1971.

130 M. A. Breuer, "A Random and an Algorithmic Technique for Fault Detection Test Generation for Sequential Circuits," *IEEE Transactions on Computers*, Vol. C-20, No. 11, p. 1364-70, November 1971.

131 M. A. Breuer, "An Algorithm for Generating a Fault Detection Test for a Class of Sequential Circuits", *Theory of Machines and Computation*, p. 313-326, 1971, Z. Kohavi and A. Paz Edrs., Academic Press, 1971.

132 M. A. Breuer, "Recent Developments in the Automated Design and Analysis of Digital Systems", IEEE, Vol. 60, No. 1. p. 12-27.

133 M. A. Breuer, "Generation of Fault Tests for Linear Logic Networks", *IEEE Transactions on Computers*, Vol. C-21, No. 1, p. 79-83, January 1972.

134 M. A. Breuer, "A Note on Three-Valued Logic Simulation", *IEEE Transactions on Computers*, Vol. C-21, No. 4, p. 399-402, April 1972.

135 M. A. Breuer, "Generation of Fault Detection Tests for Intermittent Faults in Sequential Circuits", 1972 International Symposium on Fault-Tolerant Computing, June 19-21, 1972, p. 53-57.

136 M. A. Breuer, "Automatic Digital Test Generation for Sequential and Combinational Logic", Proceedings National Nepcon, 1972, p. 109-117.

137 M. A. Breuer, *Design Automation of Digital Systems*, Prentice Hall, 1972.

138 M. A. Breuer, "Testing for Intermittent Faults in Digital Circuits", *IEEE Tranactions on Computers*, Vol. C-22, No. 3, pp. 241-246, March 1973.

139 M. A. Breuer and Loyd Harrison, "Procedures for Eliminating Static and Dynamic Hazards in Test Generation", Electrical Engineering Department University of Southern California, Los Angeles, March 28, 1973, also IEEE Computer Society Repository, R73-251.

140 M. A. Breuer, "The Effects of Races, Delays and Delay Fault on Test Generation", IEEE Computer Society Repository R73-290.

141 A. Brown and H. Young, "Toward and Algebraical Theory of the Analysis and Testing of Digital Networks", Session VII G4, 15th Annual Astronautical Society, 17-20 June 1969, Denver, Colorado.

142 A. Brown, D.D. Johnson, and M. Young, "An Algebraic Theory of the Analysis and Testing of Digital Networks", 35th National Meeting of Operations Research, 1969.

143 A. Brown and H. W. Young, "Algebraic Logic Network Analysis (ALNA) Toward an Algebraic Theory of the Analysis and Testing of Digital Networks", IBM Systems Development Division, Report Tr. 001974, 19 January 1970.

144 A. Brown, "An Algebraic Approach to Test Data Generation", Proceedings of the ACM Annual Conference, 1972, p. 352-367.

145 J. R. Brown, "Pattern Sensitivities in Mos Memories," Proceedings of the Symposium-Testing to Integrate Semiconductor Memories Into Computer Mainframes, October 4, 1972.

146 H. Brownstein, "Sequential and Nonsequential Detection Procedures, Based on Order Statistics," Ph. D. Dissertation, NYU, 1969.

147 J. D. Brule, R.A. Johnson, and E. Kletsky, "Diagnosis of Equipment Failures", *IRE Transactions on Reliability and Quality Control* Vol. RQC-9, p. 23-34, April 1960.

148 L. Buchsbaum, *et al.*, "Investigation of Fault Diagnosis by Computational Methods", UNIVAC. Blue Bell, Pennsylvania, AD-601 204, 1964.

149 D. A. Bukhakova, "Construction of Tests for Logical Devices with Memory by the Method of Essential Paths", *Automation and Remote Control*, Vol. 32, No. 10, p. 1680-1685, October 1971.

150 D. C. Burnstine and W. H. Eppard, "Maintenance Strategy Diagramming Technique", Proc. 12th Nat. Symp. Rel. and Quality Control, p. 497-506, 1966.

151 R. Butterworth, "A Branch and Bound Method for Optimal Fault Finding", Ph.D. Ph.D. Thesis, University of California, August 1969.

152 R. Butterworth, "Some Reliability Fault-Testing Models," *Operations Research*, Vol. 20, No. 2, p. 335-43, March-April 1972.

153 B. B. Buyanov, S. M. Domanitskiy, and V. N. Ozernoi, "Tests for Logical Systems of Unifunctional Elements, *Engineering Cybernetics*, Vol. No. 2, p. 83-96, 1966.

154 J. G. Calderone, G.D. Kraft, and S. F. Samson, "Computer Controlled Memory Test Processor", Proceedings NEC 1970, Vol. 26, p. 63-64.

155 R. C. Calhoun, "Diagnostics at the Microprogram Level", *Modern Data Systems*, No. 5, p. 58, 1969.

156 E. L. Calvin, "Computer Program Detects Transient Malfunctions in Switching Circuits", NASA TECBRIEF 67-10002, January 1967.

157 A. B. Carroll, M. Kato, Y. Koga, and K. Naemura, "A Method of Diagnostic Test Generation", SJCC 1969, p. 221-8.

158 B. D. Carroll, "A Tabular Method for Generating Test Sequences for Sequential Networks", Proceedings of the 5th Hawaii International Conference on System Science, Honolulu, Hawaii 11-13, January 1972, p. 290-2.

159 B. D. Carroll and E. W. Smith, "A Bibliography of Fault-Tolerant Computing", AD-739522, February 1972.

160 B. D. Carroll and D. M. Jones, "Test Sequences for Sequential Logic Circuits", IEEE Region III Convention, 1973.

161 C. C. Carroll, "Efficient Generation of Minimum Fault Test Schedules for Combinational Logic Networks", Digital Systems Laboratory Technical Report No. AU-T-19, Auburn University, Auburn, Alabama, August 1971.

162 C. C. Carroll and L. Beranson, "Design of Minimum Fault Test Schedules and Testable Realizations for Combinational Logic Circuits", Project Themis Technical Report AU-T-8.

163 W. C. Carter, H. C. Montgomery, R. J. Preiss, and H. J. Reinheimer, "Design of Serviceability Features for the IBM System/360, *IBM Journal on Research and Development*, Vol. 8, No. 2, p. 115-26, April 1964.

164 P. Castello, *"Comprehension Et Depannage Rapide Des Reseaux Combinatoires Et Des Systemes A Sequences,"* GAMI-ISMCM, Colloque 1969, 16 June 1969, *Mecanique Et Electricite*, Vol. 243, March 1970, p. 25-28.

165 H. Y. Chang, "An Algorithm for Selecting an Optimum Set of Diagnostic Tests",

IEEE Transactions on Electronic Computers, Vol. EC-14, No. 5, p. 706-711, October 1965.

166 H. Y. Chang and W. Thomis, "Methods of Interpreting Diagnostic Data for Locating Faults in Digital Machines", *BSTJ*, Vol. 56, p. 289-317, February 1967.

167 H. Y. Chang, "A Method for Digitally Simulating Shorted Input Diode Failures", IEEE Computer Group Repository R-68-69, December 11, 1967.

168 H. Y. Chang, "A Distinguishability Criterion for Selecting Efficient Diagnostic Tests", SJCC 1968, p. 529-34.

169 H. Y. Chang, "Figures of Merit for the Diagnostics of a Digital System", *IEEE Transactions on Reliability*, Vol. R-17, No. 3, p. 147-53, September 1968.

170 H. Y. Chang, "Techniques for Diagnosing Faults in Switching Systems", Proc. Symposium on Information Processing, Pages 60-72, Purdue University, April 1969.

171 H. Y. Chang, "A Method for Digitally Simulating Shorted Input Diode Failures", *BSTJ*, August 1969, p. 1957-66.

172 H. Y. Chang, E. Manning, "Logic Analyzer and Maintenance Planning an Overview", IEEE Cornell Conference 1969, Vol. 2, pp. 260-81.

173 H. Y. Chang and J. M. Scanlon, "Design Principle for Processor Maintainability in Real-Time Systems", FJCC, 1969, p. 319-28.

174 H. Y. Chang, R. C. Dorr, and R. A. Elliott, "On Simulator Use in Designing Self-Checked Logic Circuits", Proceedings of the Open Workshop on Fault Detection and Diagnosis in Digital Circuits and Systems, Lehigh University, p. 11, 7-9 December 1970.

175 H. Y. Chang, E. Manning, and G. Metze, *Fault Diagnosis of Digital Systems*. J. Wiley, 1970.

176 H. Y. Chang, R. C. Dorr, and R. A. Elliott, "Logic Simulation and Fault Analysis of a Self-Checking Switching Processor", In Proc. 1972 IEEE-ACM Design Automation Workshop, June 1972, pp. 128-135.

177 H. Y. Chang, R. C. Dorr, and R. A. Elliott, "Hardware Maintainability and Software Reliability in Electronic Switching Systems", 10th Annual Conference on Circuit and System Theory, October 4-6, 1972.

178 S. J. Chang, S. Y. H. Chu, and M. A. Breuer, "Detection and Location of Multiple Stuck-Type Failures in Sequential Circuits", IEEE Computer Society Repository, R-72-223.

179 S. Chappell, H. Chang, C. Elmendorf, and L. Schmidt, "Comparison of Parallel and Deductive Fault Simulation Methods", Workshop on Diagnosis and Reliable Design of Digital Systems, Pasadena, December 5-7, 1973.

180 S. Chappell and S. S. Yau, "Simulation of Large Asynchronous Logic Circuits Using Using an Ambiguous Gate Model", FJCC, 1971, p. 651-661.

181 G. G. Charaev, "Testing of Operative State and Fault Diagnosis in an Incompletely Homogeneous Two-Dimensional Structure", *Automation and Remote Control*, July 1968, p. 1130-1137.

182 G. G. Charaev, "Fault Diagnosis for Logic Devices Using Three Input Elements", *Engineering Cybernetics*, Vol. 8, No. 4, p. 718-23, July 1970.

183 G. G. Charaev, "Technical Diagnostics of Threshold Element Circuits", *Automation and Remote Control*, Vol. 32, No. 11, p. 1832-37, November 1971.

184 I. A. Chegis and S. Y. Yablonskiy, "Logical Methods of Checking the Operation of Electrical Circuits", Trudy Matem. Inst. Im V. A. Steklova, IZD. An SSR, No. 51, p. 270-360, 1958, (Transl. JPRS-5173, OTS-6051069).

185 P. O. Chelson, "Program Listing for Fault Free Analysis of JPL, "Technical Report 32-1542, Report NASA-CR-125064, December 1971.

186 D. D. Cheng, "A Log-Out Analysis Method for Fault Detection in Computer", International Symposium on Fault-Tolerant Computing, Pasadena, California, 1-3 March 1971, p. 105-107.

187 G. Chernow, "Boosting Plated-Wire Yield-Which Knob to Adjust," Electronics, September 1, 1969, pp. 95-96.

188 D. K. Chia,"Structural Distinguishability of Faults in Combinational Networks," MS Thesis, University of Wisconsin, 1969.

189 D. K. Chia, S. J. Hong and M. Y. Hsiao, "Determination of Redundant Elements in Combinational Logic Networks, IBM TR. 002070," 29 July 70, Poughkeepsie, N. Y., and IEEE Computer Society Repository, R-71-12.

190 D. K. Chia and M. Y. Hsiao, "An Experimental Program for Asynchronous Sequential Circuit Testing Based on Boolean Difference, Open Workshop on Fault Detection and Diagnosis in Digital Systems", 6-8 December 1971, Lehigh University.

191 D. K. Chia and M. Y. Hsiao, "A Homing Sequence Generation Algorithm for Fault Detection in Asynchronous Sequential Circuits, 1972" International Symposium on Fault-Tolerant Computing, 19-21 June 1972, p. 137-142.

192 D. K. Chia, J. W. Cho, M. Y. Hsioao, and A. M. Patel, "Algorithm for Generating a Complete Minimum Set of Test Pattern for Combinational Circuits", IBM Technical Disclosure Bulletin, Vol. 14, March 1973, pp. 3060-3063.

193 A. C. L. Chiang, I. S. Reed and A. V. Banes, "Path Sensitization Partial Boolean Difference, and Automated Fault Diagnosis," IEEE Transactions on Computers, Vol. C-21, No. 2, p. 189-195, February 1972.

194 C. Chicoix, C. Durante, J. Erceau, and J. P. Richard, "Influence de La Topologie sur La Detection Des Pannes Dans Les Circuits Combinatoires", Colloque Systemes Logiques, Bruxelles, 1969.

195 C. Chicoix, J. Erceau, F. Blanca, and P. Prechoux, "Dialog Programme D'Eleaboration De Sequences De Test De Detection Et De Diagnostic Des Pannes Logiques Dans Les Multipoles Combinatoires", Congres De La Microelectronique Avancee, Paris, Avril 1970.

196 C. Chicoix and J. Nanard, "Methode De Recherche Des Tests De Detection Dans Les Multipoles Combinatoires," "Applications A La Recherche De La Validite D'Une Sequence De Test Donnee", Colloque International Conception Et Maintenance Des Automatismes Logiques, Toulouse, September 27-28, 1972, also N73-12256.

197 V. P. Chipulis, "On the Construction of Tests for Checking Combinational Circuits", Automation and Remote Control, October 1970.

198 V. P. Chipulis, "Construction of Complete Checking Tests," Automation and Remote Control, Vol. 32, No. 11, p. 1838-42, November 1971.

199 J. W. Cho and P. N. Kothari, "Automated Test Pattern Generation Algorithm Using Boolean Difference Technique for Combinational Logic Networks", IBM Technical Disclosure Bulletin, Vol. 15, March 1973, pp. 3014-3015.

200 W. L. Chu, "Adpative Diagnosis of Faulty Testing", *Journal of Operations Research*, Vol. 16, p. 915-27, 1968.

201 W. W. Chu, "Diagnosis of System Failures", IBM Tech Rept. TR-02294, March 26, 1964.

202 W. W. Chu, "A Mathematical Model for Diagnosing System Failures", *IEEE Transactions on Electronic Computers*, Vol. EC-16, No. 3, p. 327-31, June 1967.

203 W. W. Chu, "Some Recent Results on Diagnosis of System Failures," XVI General Assembly of International Scientific Radio Union (URSI), Ottawa, August 1969.

204 Y. H. Chuang and S. R. Vishnubhotla, "A Path Analysis Approach to the Diagnosis of Combinational Circuits", 8th Design Automation Workshop, 28-30 June 1971.

205 G. P. Chubb and R. G. Mills, "Development and Preparation of Cost-Optimized Troubleshooting Decision Trees", Proceedings of the IEEE Conference on Maintainability, November 1969, p. 119-129, and AD-731 145, November 1969.

206 L. Ya Chumakov, "Faults of One System of Computer Elements", In *Computer Fault Diagnosis*, IZD, Nauka 1965.

207 G. Cioffi and L. Fiorillo, "Diagnosis and Utilization of Faulty Universal Tree Circuits", SJCC, 1969, p. 139-147.

208 B. W. Clark, "Automatic Fault Location of Processing Units," Proc. Auto. Support Symp. for Advanced Maintainability, 1967.

209 F. W. Clegg, "The Spoof, A New Technique for Analyzing the Effects of Faults on Logic Networks", IEEE Computer Society Repository, R-71-105, and Digital Systems Laboratory, Rept. No. 11, Stanford University, 1970.

210 F. W. Clegg and E. J. McCluskey, "Algebraic Properties of Faults in Logic Networks", IEEE Computer Society Repository, R 71-106, and Digital Systems Laboratory Technical Report No. 4, Stanford Electronics Laboratories Report No. SU-SEL-69-078, March 1970.

211 F. W. Clegg, "Algebraic Properties of Faults in Logic Networks", Ph.D. Dissertation, Dept. of Electrical Engineering, Stanford University, Stanford, California, June 1970.

212 F. W. Clegg, "The Algebraic Approach to Faulty Logic Networks", Proceedings of the International Symposium on Fault-Tolerant Computing, 1-3 March 1971, Pasadena, California, p. 44-45.

213 F. W. Clegg, "Use of Spoofs for Faulty Logic Network Analysis", 1972 International Symposium on Fault-Tolerant Computing, 19-21 June 1972, p. 143-47.

214 F. W. Clegg, "Use of Spoof's in the Analysis of Faulty Logic Networks", *IEEE Transactions on Computers*, Vol. C-22, No. 3, pp. 229-234, March 1973.

215 D. J. Cohen and E. G. Manning, "Fault Simulator for Research Applications," Research Report CSRR 2012, Department of Applied Analysis and Computer Science, University of Waterloo, 1969.

216 D. J. Cohen and E. G. Manning, "A Fault Simulator for Research Applications", IEEE Computer Society Repository, R-69-142, July 18, 1969.

217 D. J. Cohen, "Computer Based Fault Analysis of Digital Systems", Research Report CSRR2020, University of Waterloo, Dept. of Applied Analysis and Computer Science, Waterloo, Canada, April 1970.

218 D. J. Cohen and E. G. Manning, "Programmes Pour Faciliter La Conception De

Circuits Integres A Grande Echele", Colloque International Sur La Micro-electronique Avancee, Paris 1970, Tome I, p. 461-70, Editions Chiron.

219 J. J. Cohen and L. A. Whittaker, "Improved Techniques in Diagnostic Program-ming", *The Sylvania Technologist*, Vol. 3, p. 90-6, July 1960.

220 J. J. Cohen and L. A. Whittaker, "An Approach to Diagnostic Programming", 3rd Annual General Meeting ACM, 24-25 August 1960.

221 M. Cohn , "Controllability in Linear Sequential Networks", *IRE Transactions on Circuit Theory*, Vol. CT-9, pp. 74-78, March 1962.

222 M. Cohn and G. Ott, "The Design of Adaptive Procedures for Fault Detection and Isolation", Sperry Rand Research Report, SRRC-RR-66-64, October 1966, Program 13-551-70.

223 M. Cohn and G. Ott, "The Design of Adaptive Procedures for Fault Detection and Isolation", *IEEE Transactions on Reliability*, Vol. R-20, No. 1, p. 7-10, February 1971.

224 F. B. Cole, "Automatic Generation of Functional Logic Test Programs Through Simulation", 7th Annual Design Automation Workshop, 1970, p. 116-127.

225 R. Colgan, "Design Consideration to Improve Testability", IEEE 1974 Intercon.

226 E. Collins, "The Fairchild System for Computer-Aided Test Design, Wescon 1973.

227 A. J. Collmeyer, "A Random Sampling Approach to Digital Detection", Ph.D. Dissertation, SMU, 1969.

228 J. N. Contensou and Y. Picot, "Systemes-Surete De Fonctionnement Et Securite", AFIRO-B, May-June 1968, No. 9.

229 J. N. Contensou, "Comparison Logique Du Diagnostic Hierarchique Et Du Diagnostic Medical", Document IRIA, January 28, 1972.

230 R. W. Cook and M. J. Flynn, "Logical Network Cost and Entropy", IEEE Computer Society Society Repository, R-71-41.

231 R. C. Cork and C. H. Salzmann, "Devising Patterns to Test Complex Logic Circuits", *Electronics*, August 14, 1972, p. 120-25.

232 R. C. Cork and C. H. Salzmann, "An Automated Approach to Test Vector Genera-tion", Compcon, 1972.

233 D. Corman and J. G. Burns, "A Comparison of Digital Simulation and Physical Fault Insertion in Diagnostic Test Development", 1972 International Symposium on Fault-Tolerant Computing, 19-21 June 1972, p. 42-46.

234 M. Correia and D. K. Ferguson, "Production Circuit Array Test System", *Digest of the IEEE Computer Conference*, p. 81-82, 1968.

235 M. Correia, D. Cossman, F. Putzolu, and T. Snethen, "Minimizing the Problem of Logic Testing by the Interaction of a Design Group with User-Oriented Facilities", Seventh Design Automation Workshop, p. 100-107, June 1970.

236 M. Courvoisier, "Contribution A La Detection Et a la Localisation des Pannes dans les Systemes Sequentiels", These 3eme cycle, Universite de Toulouse, 14 January 1971.

237 M. Courvoisier and M. Diaz, "Programme D'Analyse Booleenne des Systemes Sequentiels", *Automatisme*, Tome XVI, No. 12, December 1971, p. 642-647.

238 M. Courvoisier, "Localisation des Pannes Simples D'Une Machine Sequentielle a Partir des Sequences de Synchronisation", *Automatisme*, Tome XVII, No. 10, p. 291-96, October 1972.

239 M. Courvoisier, "Elaboration de Tests de Detection de Pannes dans les Systemes Sequentiels", *Automatisme*, Vol. XVII, No. 10, p. 297-304, October 1972.

240 K. J. Crooks and J. Blythin, "A Computer Controlled Tester for Logic Networks and a Method for Synthesizing Test Patterns," *Radio Electronic Engineer*, Vol. 40, No. 6, p. 309-15, December 1970.

241 C. E. Cunningham and W. Cox, *Applied Maintainability*, John Wiley, 1972.

242 N. L. Daggett and E. S. Rich, "Diagnostic Programs and Marginal Checking in Whirlwind I Computer," IRE National Convention Record, Vol. 1, Pt. 7, p. 48-54, 1953.

243 J. Danda, "An Application of a Digital Computer for Extended Shmoo-Plotting of Ferrite-Core Memories", Proceedings IFIP 1968, p. 775-77.

244 R. Dandapani, S. M. Reddy, and J. P. Robinson, "An Investigation into Redundancy and Testability of Combinational Logic Networks", AD-714 157, September 1970.

245 R. Dandapani, S. M. Reddy and J. P. Robinson, "Testable Design of Logic Networks", IEEE Computer Society Repository, R-71-63.

246 R. Dandapani, "Derivation of Minimal Test Sets for Monotonic Logic Circuits", *IEEE Transactions on Computers*, Vol. C-22, No. 7, pp. 657-661, July 1973.

247 R. Danklefs, "Testing MOS the How and Why," *Electronic Engineering*, Vol. 29, No. 10, p. 41-46, October, 1970.

248 P. Das and D. E. Farmer, "Fault Detection Experiments on Sequential Machines", Computer Science Conference, Columbus, Ohio, February 20-22, 1973.

249 P. Das and D. E. Farmer, "Detection Experiments for Parallel and Serial Decomposition of Sequential Machines", IEEE Computer Society Repository, R73-238.

250 P. S. Dauber, "An Analysis of Error in Finite Automata", *Information and Control*, Vol. 8, p. 295-303, June 1965.

251 N. I. Davidovskaya and I. S. Neishtadt, "Selection of Tests for Combinational Logic Circuits", Materials of the Conference on New Electronic Devices, Moscow, IZD. Mosk. Doma Nauchno - Tekhnicheskoi Propagandy Im 1966.

252 M. Davio and P. Piret, "Les Derivees Booleennes et Leur Application au Diagnostic", *Revue MBLE*, Vol. XII, No. 3. p. 63-76, 1969.

253 R. F. Davis, "Fault Testing in Combinational and Sequential Tree Circuits", Washington State University, Pullman, USA, Thesis, 1972.

254 E. De Atley, "LSI Testing is a Large Scale Headache", *Electronic Design*, 2 August 1969, p. 24-26, 28-32, 34.

255 A. Deb, "Exclusive Path Sensitization in a Combinational Network of Arbitrary Complexity and with Multiple Faults", Computer Science Conference Columbus, Ohio, February 20-22, 1973.

256 O. S. Denisenko, "Checking Functions for the Correct Operation of Memory Elements, in Problems of Synthesis of Finite Automata", V. I. Levin Ed., p. 131-42, Riga, USSR-IZDATEL Stvo 'Zinatne', 1972.

257 J. J. Dent, "Diagnostic Engineering", *IEEE Spectrum*, Vol. 4, p. 99-104, July 1967.

258 J. J. Dent, "Diagnostic Engineering Requirements", SJCC 1968, p. 503-507.

259 J. J. Dent, "Testing and Diagnosing—A Practitioner's Point of View", International Symposium on Fault-tolerant Computing, Pasadena, March 1-3, 1971, p. 53-55.

260 N. Deo, "The Self-Diagnosability of Computer", *IEEE Transactions on Electronic Computers*, Vol. EC-15, p. 799, October 1966.

261 P. Deschizeaux, "Applications a L'Algebre de Boole de la Notion de Derivee", Seminaire de Logique, Imag, Seminaire de Logique, February 1965.

262 P. Desmarais and M. Krieger, "Selective Testing-A New Approach to Fault Detection in Logic Circuits", IEEE Computer Society, Repository R73-213.

263 M. J. Devaney and G. W. Zobrist, "Fault Diagnosis of Operational Synchronous Digital Systems", N.70-16910, 1969, and National Meeting of Operations Research 1969.

264 M. J. Devaney, R. A. Waid, R. W. Leavene, and G. V. Lago, "Fault Diagnosis and Self Repair in Synchronous Digital Systems", Computer Science Conference, Columbus, Ohio, February 20-22, 1973.

265 A. K. Dewdney and A. L. Szilard, "Tours in Machines and Digraphs", *IEEE Transactions on Computers*, Vol. C-22, No. 7, pp. 635-39, July 1973.

266 E. Diatcu and D. N. Dobrescu, "Deux Methodes et Quatre Programmes Automatiques de Calcul de la Fiabilite des Circuits", *Revue Riro*, R3, p. 45-69, 1970.

267 M. Diaz, M. Courvoisier, et J. P. Richard, "Une Approche a la Detection en Ligne des Pannes Dans les Systemes Sequentiels Synchrones", Colloque Conception et Maintenance des Automatismes Logiques, Toulouse, September 27-28, 1972, also N73-12197.

268 R. J. Diephuis, "Fault Analysis for Combinational Logic Networks", Ph.D. Dissertation, MIT, Department of Electrical Engineering, September 1969.

269 R. J. Diephuis, "Single S-A-0, S-A-1 Fault Location Test Sets for Combinational Logic Networks", Open Workshop on Fault Detection and Diagnosis in Digital Circuits and Systems, Lehigh University, 7-9 December 1970, p. 1-7.

270 H. Dirilten, "On the Mathematical Models Characterizing Faulty Four-Phase MOS Logic Arrays", *IEEE Transactions on Computers*, Vol. C-21, No. 3, p. 301-5, March 1972.

271 J. E. Doucet and G. Gelis, "Detection et Diagnostic Automatique des Pannes dans les Circuits Logiques," Contrat DGRST 69-01-815, Universite de Toulouse, January 1971.

272 R. W. Downing, J. S. Nowak, and L. S. Tuomenoksa, "No. 1 ess Maintenance Plan", *BSTJ*, Vol. 43, No. 5, Pt. 1, p. 1961-2019, September 1964.

273 M. W. Du, "Multiple Fault Detection in Combinational Circuits", Ph. D. Dissertation, Dept. of Comput. Sci., The John Hopkins Univ., Baltimore, Md., 1972.

274 M. W. Du and C. D. Weiss, "Circuit Structure and Switching Function Verification", Allerton Conference 1972, pp. 122-130.

275 M. W. Du and C. D. Weiss, "Multiple Fault Detection in Combinational Circuits—Algorithms and Experimental Results", 1972 International Symposium on Fault-Tolerant Computing, 19-21 June 1972, p. 120-125.

276 M. W. Du and C. D. Weiss, "Multiple Fault Detection in Combinational Circuits—Algorithms and Computational Results", *IEEE Transactions on Computers*, Vol. C-22, No. 3, pp. 235-240, March 1973.

277 M. W. Du and C. D. Weiss, "Circuit Structure and Switching Function Verification", *IEEE Transactions on Computers*, Vol. C-22, No. 7, pp. 618-625, June 1973.

278 L. S. Dubitskaia and I. V. Kogan, "Algorithm for the Synthesis of a Minimized Diagnostic Test for a Multi-Output Logical Device", *Automation and Remote Control*, Vol. 33, No. 2, Pt. 2, p. 320-1, February 1972.

279 K. A. Duke, G. R. Putzolu, and J. P. Roth, "A Heuristic Algorithm for Sequential Circuit Diagnosis", IBM Research Report RC-2639, September 29, 1969.

280 C. Durante, J. Erceau, and J. P. Richard, "Sur la Detection des Pannes dans les Ensembles Logiques."

281 C. Durante, J. Erceau, C. Chicoix, and J. P. Richard, "Influence de la Topologie sur la Detection des Pannes dans les Circuits Combinatoires", Colloque sur les Systemes Logiques, p. 603-614, Bruxelles, September 1969.

282 C. Durante, "Recherches Theoriques et Techniques sur la Detection et la Localisation des Pannes dans les Sous-Ensembles Logiques", Rapport Final DGRST No. 6901715, March 1971.

283 T. F. Dwyer, "Comments on Faulty Testing and Diagnosis in Combinational Digital Circuits", *IEEE Transactions on Computers*, Vol. C-18, No. 8, p. 760, August 1969.

284 G. Edge, "System 250 Diagnostics", Proceedings of the Conference Computers-Systems and Technology, London, 24-27 October, pp. 201-212.

285 J. F. Egler, "A Procedure for Converting Logic Table Conditions into an Efficient Sequence of Test Instructions", *Communications of the ACM*, Vol. 6, No. 8, p. 510-14, September 1963.

286 E. B. Eichelberger, "Hazard Detection in Combinational and Sequential Circuits", *IBM Journal of Research and Development*, Vol. 9, No. 2, p. 90-99, 1965.

287 R. D. Eldred, "Test Routines Based on Symbolic Logical Statements," *Journal of the ACM*, Vol. 6, No. 11 p. 33-36, January 1959.

288 R. D. Eldred, "Effect of Microelectronics on Error Diagnosis", *Computer Design*, Vol. 6, No. 11, p. 6-8, 1967.

289 C. R. Elles and E. A. Dance, "Exercising Memory Systems with Worst-Case Bit Patterns", *Electronics*, September 1, 1969, pp. 93-94.

290 J. M. Elmore and P. G. Guenther, "Automatic Fault Isolation for Digital Assemblies", *IBM Technical Disclosure Bulletin*, Vol. 14, No. 12, p. 3767-70, May 1972.

291 H. Endou and J. Kawasakai, "Improved Testing Method for Wirings in Complex Hybrid Microcircuits, Applying the Grouping Technique", Proceeding of the IEEE Conference on Electronic Components, 1971.

292 P. H. Enslow and A. B. Salisbury, "A Pragmatic First Look at Diagnostic Programming for Digital Computers", AD-652-370, May 1967.

293 D. L. Epley, "Systematic Analysis of Combinational Network Hazards and Faults from Dual Matrices", Proceedings of the 2nd Annual Princeton Conference on Information Sciences and Systems, March 1968, p. 64-68.

294 J. Erceau, "Development Non Symetrique et Disjonctif D'Une Fonction Booleenne," Comptes Rendus a L'Academie des Sciences T.272, p. 1750-52, 28 June 1971.

295 J. Erceau, "Location of Single Faults in Combinational Circuits", IEEE Computer Society, R73-75.

296 L. M. Ericsson, "Computer Fault Location", Patent U. K. 1166057, 16 November 1966.

297 V. A. Ermilov, "Methods for Selecting Essential Failures for Diagnosis of Digital Circuits, I. General Expressions for Failures Possible During an Experiment," *Automation and Remote Control*, Vol. 32, No. 1, p. 141-49, January 1971.

298 V. A. Ermilov, "Construction of Minimal Diagnostic Tests of Digital Circuit with Memory by the Truncated-Tree Method", *Automation and Remote Control*, Vol. 33, No. 44, p. 679-85, April 1972.

299 W. Escher, "Fehlerdiagnose an Schaltnetzen und Schaltwerken", Third Annual Symposium of the Gesellschat fur Informatik, October 8-10, 1973.

300 A.Fadini, "Operatori che Estendono alle Algebre di Boole la Nozione di Derivata", Mat. Ital., 89, p. 42-64, 1961.

301 G. Fantauzzi, "Diagnosi Delle Catene di Maitra e Teoria dei Semigruppi", *Calcolo*, Vol. 7, Fasc. 1-2, January-June 1970.

302 G. Fantauzzi and A. Marsella, "Multiple-Fault Detection and Location in Fan-Out Free Combinational Circuits", *IEEE Transactions on Computers*, Vol. C-23, No. 1, pp. 48-55, January 1974.

303 G. Fantauzzi and A. Marsella, "Multiple Fault Detection and Location in Fan-Out Free Combinational Circuits", IEEE Computer Society Repository, R73-40.

304 D. E. Farmer, "A Strategy for Detecting Faults in Sequential Machines Not Possessing Distinguishing Sequences", *FJCC*, 1970, p. 493-501.

305 D. E. Farmer, "Fault-Detection Experiments for Sequential Machines", Ph.D. Dissertation, Polytechnic Institute of Brooklyn, 1970.

306 D. E. Farmer, "Algorithms for Designing Fault-Detection Experiments for Sequential Machines", *IEEE Transactions on Computers*, Vol. C-22, No. 2, pp. 159-167, February 1973.

307 J. L. Fike, "The Failure Modes of Large Arrays, Their Logical Representation, and the Development of Procedures Associated with Faults in Digital Equipment", Workshop on Hardware Faults and the Programmer, Palm Springs, 28-29 June 1968.

308 J. L. Fike, "Heuristic and Adaptive Techniques for Diagnostic Test Generation", Ph.D., Southern Methodist University, Dallas, Texas, December 1972.

309 S. I. Firstman and B. Gluss, "Optimum Search Routines for Automatic Fault Location", Rand Corp. and AD-616 415, 11 December 1959.

310 S. I. Firstman and B. Gluss, "Search Rules for Automatic Fault Location", Rand Corporation, RM-2514, June 1960.

311 S. I. Firstman and B. J. Voosen, *Optimizing a Pre-Launch Check-Out in Electronic Maintainability*, F. L. Ankenbrandt, Editor, Vol. 3, Engineering Publisher, Elizabeth, N. J., 1960.

312 S. I. Firstman and B. Gluss, "Optimum Search Routines for Automatic Fault Location", *Operations Research*, Vol. 8, p. 512-23, 1960.

313 M. Fischler, "The Prognosis for Design Verification and Testing of Computing Hardware and Software," Workshop on Diagnosis and Reliable Design of Digital Systems, Pasadena, California, December 5-7, 1973.

314 J. Fitzpatrick, "Today's Three Test Environments," Proceedings of the Symposium-Testing to Integrate Semiconductor Memories into Computer Mainframes, October 4, 1972.

315 M. J. Flomenhoft, "A System of Computer Aids for Designing Logic Circuit Tests", 7th Annual Design Automation Workshop, 1970, p. 128-131.

316 M. J. Flomenhoft, S. C. Si, and A. K. Susskind, "Algebraic Techniques for Several Fault Types", 1973 International Symposium on Fault-Tolerant Computing, Palo Alto, California, June 20-22, 1973.

317 M. J. Flynn and S. A. Szygenda, "A Perspective on Computer Reliability", Proceedings of the 1972 Annual Reliabilty and Maintainability Symposium, p. 166-173.

318 M. E. Fohl, "Central Processor Diagnosis by Functions", 1972 International Symposium on Fault-Tolerant Computing, 19-21 June 1972, p. 62-67.

319 R. E. Forbes, C. B. Stieglitz, and D. Muller, "Automatic Fault Diagnosis", AIEE Conference on Diagnosis and Failures in Switching Circuits, 15-16 May 1961.

320 K. A. Foster, "Comments on Algorithms for Designing Fault-Detection Experiments for Sequential Machines", IEEE Computer Society Repository, R73-154.

321 F. Fowler, "Software Aids to Diagnostic Programming", 1971 IEEE Systems, Man and Cybernetics Group Annual Symposium Record, Anaheim, California, 25-27 October 1971, p. 366.

322 F. Fowler, "Salvaging Diagnostic Fault Dictionaries and Fault Tallies", Compcon 1972, p. 251-54.

323 M. Fridrich and W. A. Davis, "Single Fault Detection in Combinational Networks", Canadian Computer Conference Session 1972, Montreal, Canada, June 1-3, 1972.

324 A. Friedes, "The Propagation of Fault Lists Through Combinational or Sequential Circuits", Open Workshop on Fault Detection in Digital Circuits and Systems, p. 12-41, 7-9 December 1970, Lehigh University.

325 A. D. Friedman, "Fault Detection in Redundant Circuits", *IEEE Transactions on Electronic Computers*. Vol. EC-16, No. 1, p. 99-100, February 1967.

326 A. D. Friedman and P. R. Menon, "Design of Generalized Double Rank and Multiple Rank Sequential Circuits", *Information and Control*, Vol. 15, p. 436-451, November 1969.

327 A. D. Friedman and P. R. Menon, "Easily Testable Combinational Circuits," Open Workshop on Fault Detection and Diagnosis in Digital Circuits and Systems, Lehigh University, p. 8, 7-9 December 1970.

328 A. D. Friedman and P. R. Menon, *Fault Detection in Digital Circuits*, Prentice Hall, 1971.

329 A. D. Friedman and P. R. Menon, "Fault Location in Iterative Logic Arrays", *Theory of Machines and Computations*, Z. Kohavi and A. Paz, p. 327-340, Academic Press, 1971.

330 A. D. Friedman and P. R. Menon, "Comments on Design of Diagnosable Iterative Arrays", *IEEE Transactions on Computers*, Vol. C-21, No. 5, p. 511, May 1972.

331 A. D. Friedman and P. R. Menon, "Restricted Checking Sequences for Sequential Machines", *IEEE Transactions on Computers*, Vol. C-22, No. 4, pp. 397-399, April 1973.

332 A. D. Friedman, "Diagnosis of Nonclassical Faults in Combinational Circuits", 1973 International Symposium on Fault-Tolerant Computing, Palo Alto, California, June 20-22, 1973.

333 A. D. Friedman, "Easily Testable Iterative Systems", *IEEE Transactions on*

Computers, Vol. C-22, No. 12, pp. 1061-1064, December 1973.

334 H. Fugiwara and K. Kinoshita, "Design of Diagnosable Sequential Machines Utilizing Extra Outputs", IEEE Computer Society Repository, R73-105.

335 T. Fukuri and Y. Nakanishi, "Determination of Input Sequence for Fault Detection in Finite State Systems", *Electronics and Communications in Japan*, Vol. 50, No. 10, p. 174-82, October 1967.

336 T. Fukuri and Y. Nakanishi, "A Method of Designing Output Sensors for Equipment Based on Information Content of Output Sensors," *Electronic and Communications in Japan*, Vol. 51-C, No. 1, p. 113-21, January 1968.

337 T. Fukuri and Y. Nakanishi, "Exchangeability Between Output Sensors and Test Inputs for Failure Diagnosis", *Electronics and Communications in Japan*, Vol. 51-C, No. 10, p. 131-2, October 1968.

338 T. G. Gaddess, "Improving the Diagnosability of Modular Combinational Logic by Test Point Insertaion", Coordinated Science Laboratory, Report R-409, University of Illinois, Urbana, March 1969.

339 J. M. Galey, "Diagnosis of Failures in Sequential Circuits", Research Memo, SR-157, IBM Corp., 24 April 1961.

340 J. M. Galey, R. E. Norby and J. P. Roth, "Techniques for the Diagnosis if Switching Circuit Failures", *IEEE Transactions on Communications and Electronics*, Vol. 83, No. 74, p. 509-14, September 1964.

341 O. N. Garcia and T. R. N. Rao, "On the Methods of Checking Logical Operations", Proceedings of the Second Annual Princeton Conference on Information Sciences and Systems, p. 89-95, 1968.

342 R. M. Garey, "Optimal Binary Decision Trees for Diagnostic Identification Problems", Ph.D. Thesis, University of Wisconsin, June 1970.

343 R. M. Garey, "Simple Binary Identification Problems", IEEE Computer Society Repository, R-71-68.

344 R. M. Garey, *Optimal Binary Identification Procedures*, 1971.

345 R. M. Garey, "Optimal Binary Identification Procedures," *Siam Journal on Applied Math*, Vol. 23, No. 2, pp. 173-186, 1972.

346 R. M. Garey, "Optimal Test Point Selection for Sequential Manufacturing Processes", *BSTJ*, Vol. 51, No. 1, p. 291-300, January 1972.

347 F. Garoche, "Theorie des Ecoulements", Seminaire Imag. 1971.

348 J. W. Gault, "The Application of Fault Indistinguishability in Combinational Networks", AD-692 420, July 1969.

349 J. W. Gault, "Test Generation Using a Network Fault Equation", Ninth Allerton Conference on Circuit and System Theory, October 1971, pp. 114-121.

350 J. W. Gault, J. P. Robinson, and S. M. Reddy, "Multiple Fault Detection in Combinational Networks", *IEEE Transactions on Computers*, Vol. C-21, No. 1, p. 31-36, January 1972.

351 J. W. Gault, "Final Report on Fault Preprocessing in Digital Networks", Electrical Engineering Department, North Carolina State University, 1972.

352 S. E. Gelenbe, "Regular Expressions and Checking Experiments", AD-666 696, September 1967.

353 S. E. Gelenbe, "The Algebraic Design of Checking Experiments", IEEE Computer Group Repository, R-69-57, 13 October 1968.

354 S. E. Gelenbe and E. J. Smith, "Two Methods for Finding Testing Experiments by the Use of Regular Expressions", Microwave Res. Inst. Programs, N68-16927 07-10, R.160-61.

355 G. Gelis, "Detection et Diagnostic des Defauts Logiques dans les Circuits Combinatoires", These de Docteur de Specialte, Universite de Toulouse, 26 October 1970.

356 J. C. Geoffroy, M. Courvoisier and M. Diaz, "Realisation de Circuits Sequentiels Asynchrones Autotestables," Rairo Serie Jaune No. J-3, Automatique 1973.

357 J. E. Gersback, "The Great Shmoo Plot-Testing Memories Automatically," Electronics, 25 July 1966, p. 127-34.

358 S. Ghosh, "Fault Detection in Arbitrarily Large Finite Cellular Arrays", AD-758-383, March 1973.

359 F. D. Giard, "Verification et Detection Automatique des Defauts de Cartes Logiques et Fiabilite", Internepcon 1973.

360 A. Gill, "State Identification Experiments in Finite State," Automata Information and Control, Vol. 4, pp. 132-154, 1961.

361 A. Gill, Introduction to the Theory of Finite-State Machines, McGraw Hill, N. Y., 1962.

362 S. Ginsburg, An Introduction to Mathematical Machine Theory, Addison-Wesley, Reading, Mass., 1962.

363 S. Ginsburg, "On the Length of the Smallest Uniform Experiment Which Distinguishes the Terminal States of a Machine", Journal of the Association of Computing Machinery, May 1958, p. 266-80.

364 E. Girard, "Programme de Reduction du Nombre de Defauts a Detector dans un Circuit Logique," Rapport LCR. DR 5, No. 1606, September 23, 1971.

365 E. Girard, J. C. Rault, and R. Tulloue, "Ecriture des Programmes de Test Automatique", Rapport Final, Avenant au Marche Cri 70027, September 1972.

366 E. Girard, J. C. Rault, and R. Tulloue, "Ecriture des Methodes de Test Probabilistes-Estimation de L'Efficacite et de la Longueur des Sequences de Test," Revue Technique Thomson-CSF, Vol., No., pp., February 1974.

367 V. V. Glagolev, "Formulation of Tests for Iterative Networks", Dokl. An. SSSR, Vol. 144, No. 6, p. 1237-40, 1962, Soviet Physics Doklady, Vol. 7, p. 480-82, 1963.

368 B. Gluss, "An Optimum Policy for Detecting a Fault in a Complex System", Operations Research, Vol. 7, p. 468-77, July-August 1959.

369 H. C. Godoy and R. E. Vogelsberg, "Single Pass Error Effect Determination," IBM Technical Disclosure Bulletin, Vol. 13, No. 11, p. 3443-3444, April 1971.

370 H. C. Godoy and R. E. Vogelsberg, "A Deductive Methods for Simulating Faults", IBM Technical Disclosure Bulletin, Vol. 13, p. 3443-44, April 1971.

371 P. Goel and D. P. Siewiorek, "Comments on Multiple Fault Detection in Combinational Networks", IEEE Computer Society Repository, R73-44.

372 W. Goerke, "An Estimate of the Necessary Length of Fault Detection Experiments for Sequential Circuits", Computer Science Conference, Columbus, Ohio, February 20-22, 1973.

373 F. M. Goetz, "Computer Aided Diagnostic Design for Electronic Switching Systems", 7th Annual Design Automation Workshop, 1970, p. 178-183.

374 Yu. A. Gogin and V. V. Kudryavtsev, "The Transfer Function of Relay-Contact Networks Having Element Failures", *Engineering Cybernetics*, No. 1, p. 59-64, 1966.

375 R. S. Goldman and V. P. Chipulis, "On the Diagnosis of Combinational Circuits Failures, Automation and Remote Control", March 1971.

376 G. A. Golubeva, D. M. Gobman, A. N. Kiseleva, and N. V. F. C. Hennie, *Finite-State Models for Logical Machines*, J. Wiley, 1968.

377 G. Gonenc, "An Algorithmic Method for Organizing Checking Experiments", Polytechnic Institute of Brooklyn, Research Report No. PIBEE 680006, June 1968.

378 G. Gonenc, "A Method for the Design of Fault Detection Experiments", *IEEE Transactions on Computers*, Vol. C-19, No. 6, p. 551-558, June 1970.

379 D. M. Goodman, "Problems and Pitfalls in Automatic Test Computer Programming", In Vol. III, *Automation in Electronic Test Equipment*.

380 B. B. Gordon, "Survey of Current Fault Isolation Techniques", Proceedings on Automatic Checkout Techniques, Battelle Memorial Institute, September 1962, p. 103-111.

381 W. Gorke, "Fault Detection in Sequential Circuits by Using Boolean Differences", *Elektron, Rechenanlagen*, Vol. 14, No. 2, p. 78-81, April 1972.

382 V. R. Gorovoi, "Synthesis of Optimal Switching Circuit with Simultaneous Construction of Test Tables", *Automation and Remote Control*, Vol. 3, No. 3, March 1968, p. 481-90.

383 V. R. Gorovoi, "On the Diagnosis of Combinational Switching Devices", *Automation and Remote Control*, November 1968.

384 Yu. M. Gorskii and V. V. Novorusskii, "A Method of Probabilistically Bounding the Credibility of Hypotheses in Complex Diagnostic Systems", *Engineering Cybernetics*, No. 5, p. 110-116, 1967.

385 E. P. Graney, "Maintenance and Acceptance Tests Used on MIDAC", *JACM*, Vol. 2, p. 95-98, April 1955.

386 J. Grason, "LSI Testing Research at Carnegie-Mellon University", Workshop on Diagnosis and Reliable Design of Digital Systems, Pasadena, December 5-7, 1973.

387 M. Gravitz, "A Note on a Modified Ternary Simulator Capable of Initializing all Fault Machine Memory Elements," *IEEE Transactions on Computers*, Vol. C-22, No. 11, pp. 1042-1044, November 1973.

388 M. Gravitz, "A Note on a Modified Ternary Simulator Capable of Initializing All Fault Machine Memory Elements", IEEE Computer Society Repository, R-72-237.

389 F. G. Gray and J. C. Meyer, "Locatability of Faults in Combinatorial Networks," *IEEE Transaction on Computers*, Vol. C-20, No. 11, p. 1047-12, November 1971.

390 C. W. Green, "Computer Controlled Test Systems for IC Memory," 1970 IEEE International Convention Record, p. 60-61.

391 G. E. Gregory and S. F. Daniels, "Final Isolation of Faults in Testing Digital Modules", International Electrical, Electronics Conference, Toronto, Canada, October 1-3, 1973.

392 S. Greibach, "Checking Automata and One Way Stack Languages," *Journal of Computer and System Sciences*, Vol. 3, p. 196-217, May 1969.

393 J. H. Griesmer, "A Method for the Diagnosis of Failures in Combinational Circuits," IBM Corp., Research Memo, SR. 158, 25 April 1961.

394 R. Griscom, "Symptom Fix File for a Microprocessor Workshop on Diagnosis and Reliable Design of Digital Systems", Pasadena, California, December 5-7, 1973.

395 D. M. Grobman, "Programmed Testing and Diagnosis of Computer Faults", IZDAT. Nauka, Moscow, 1965.

396 I. S. Grunskii and Yu. A. Rubanovich, "Algorithm for Constructing Output Test Points," *Automation and Remote Control*, Vol. 32, No. 11, p. 1826-31, November 1971.

397 I. S. Grunskii, "Checking Experiment with Strongly-Connected Diagnosed Automation", *Automation and Remote Control*, Vol. 33, No. 3, Pt. 2, p. 470-3, March 1972.

398 R. F. Guarino, "Digital Logic Fault Isolation Simulation," M.S. Thesis, Department of Electrical Engineering, Polytechnic Institute of Brooklyn, June 1970.

399 C. Guerin, "Test des Systemes Sequentiels", Ensimag, August 1973.

400 R. M. Guffin, "Microdiagnostics for Standard Computer MLP-900 Processor", *IEEE Transactions on Computers*, Vol. C-20, No. 7, p. 803-808, July 1971.

401 D. Guinet and C. Levy, "La Simulation Numerique Pour L'Analyse de la Securite et de la Maintenance des Ensembles Logiques", Colloque International Conception et Maintenance des Automatismes Logiques, Toulouse, September 27-28, 1972; also N73-12199.

402 R. Haas, "Automatic Fault Location Using On-Line Simulation," IEEE 1974 Intercon.

403 M. B. Hack, "Functional Test Philosophy for a Limited Funded Digital Computer", *Computer Design*, Vol. 9, No. 4, p. 179-183, April 1970.

404 F. J. Hackl and R. W. Shirk, "An Integrated Approach to Automated Computer Maintenance", IEEE Conference Record on Switching Circuit Theory and Logical Design, p. 289-300, 6-8 October 1965.

405 J. Hadjilogiou and B. Patel, "Design of Minimum Number of Fault Tests for Combinational Switching Circuits," Master Thesis at Florida Institute of Technology, June 1972.

406 J. Hadjilogiou and B. Patel, "Design of Minimum Number of Fault Tests for Combinational Switching Circuits," Proceedings of the IEEE Region III Convention 1973.

407 F. Hadlock, "On Finding a Minimal Set of Diagnostic Tests," *IEEE Transactions on Electronic Computers*, Vol. EC-16, No. 5, p. 674-75, October 1967.

408 J. R. Hahn, "A Maintenance Approach for a Large Computer System," Lehigh University, Open Workshop on Fault Detection and Diagnosis in Digital System, 6-8 December 1971.

409 S. L. Hakimi and A. T. Amin, "Characterization of Connection Assignment of Diagnosable Systems", *IEEE Transactions on Computers*, Vol. C-23, No. 1, pp. 86-88, January 1974.

410 T. Hallin and J. Quinn, "Design Verification Using Digital Simulation", Workshop on Diagnosis and Reliable Design of Digital Systems, Pasadena, December 5-7, 1973.

411 J. R. Hanne and R. M. Jennings, "User Assurances in LSI Descretionary Routing," Digest of Technical Papers,1968 Computer Group Conference, p. 27-28.

412 J. R. Hanne, "CAD and Testing of Discretionary Wired LSI," 1968 Computer Group Conference Digest, p. 27-28.

413 J. R. Hanne, "Computer-Aids Speed Discretionary Wiring," *Electronic Design*, P. C. 10-C, 16, 21 November 1968.

414 J. M. Hannigan and C. G. Masters,"Redundant System Test Point and Mission Reliability Procedures", *IRE Transactions on Electronic Computer*, Vol. EC-16, No. 5, p. 591-96, December 1967.

415 W. J. Happ, "Symbol-Oriented Approach to Automated Failure Analysis", 35th National Meeting of Operations Research, 1969.

416 F. H. Hardie and R. J. Suhocki, "Design and Use of Fault Simulation for Saturn Computer Design", Wescon Technical Paper, Vol. 10, Pt. 4, pp. 1-22, 1966.

417 F. H. Hardie and R. J. Suhocki, "Design and Use of Fault Simulation for Saturn Computer Design", *IEEE Transactions on EC*, Vol. EC-16, No. 4, p. 412-29, August 67.

418 P. A. Harding and M. W. Roland, "Bit Access Problems in 2 ½D 2 Wire Memories," 1967, *FJCC*, pp. 353-362.

419 L. M. Hardy, *et al.*, "Using the Huffman Code for Sequential Diagnosis," IEEE Transactions on Systems, Man, and Cybernetics, October 1971, p. 389-391.

420 C. Harlow and C. L. Coates, "Inessential Errors in Sequential Machines," IEEE Transactions on Computers, Vol. C-20, No. 6, p. 688-690, June 1971.

421 D. R. Harper and D. S. Jones, "Concurrent Memory Diagnostics," IBM Tech. Disclosure Bulletin, Vol. 14, No. 8, p. 2366-7, January 1972.

422 M. A. Harrison, "Introduction to Switching and Automata Theory, " McGraw Hill, New York, 1965.

423 M. A. Harrison, "An Analysis of Errors in Finite Automata, Information and Control," Vol. 8, p. 430-50, August 1965.

424 D. W. Hartman, "An Algorithm for Fault Detection in Synchronous Sequential Circuits," Master's Thesis, Department of Electrical Engineering, MIT, June 1968.

425 F. B. Hartman, "Boolean Differential Calculus", IBM TR.22-256, 20 December 1967.

426 J. Hartmanis and R. E. Stearns, "A Study of Feedback and Errors in Sequential Machines," IEEE Transactions on Electronic Computers, Vol. C-12, p. 233-32, June 1963.

427 J. Hartmanis and R. E. Stearns, "Algebraic Theory of Sequential Machines", Prentice Hall, 1966.

428 A. Hashimoto, T. Kasami, and H. Ozakj, "A Study of Fault Diagnosis of Logical Networks", Computer Studies, 1, 1963.

429 A. Hashimoto, T. Kasami, and H. Ozaki, "Fault Diagnosis of Logical Networks," Electronics and Communications in Japan, Vol. 47, No. 4, p. 237-49, April 1964.

430 T. Hayashi, "FACOM 230 Series Computer Design Automation System", 1970 Share-ACM -IEEE Design Automation Workshop. p. 230-42.

431 J. P. Hayes, "A Study of Digital Network Structure and Its Relation to Fault

Diagnosis", Coordinated Science Laboratory Report R-467, University of Illinois, Urbana, May 1970, also AD-707 691.

432 J. P. Hayes, "A NAND Model for Fault Diagnosis in Combinational Logic Networks", IEEE Transactions on Computers, Vol. C-20, No. 12, p. 1496-1506, December 1971.

433 J. P. Hayes, "On Realizations of Boolean Functions Requiring a Minimal or Near Minimal Number of Tests", *IEEE Transactions on Computers*, Vol. C-20, No. 12, p. 1506-13, December 1971.

434 J. P. Hayes and A. D. Friedman, "Test Point Placement to Simplify Fault Detection," 1973 International Symposium on Fault-Tolerant Computing, Palo Alto, California, June 20-22, 1973.

435 J. P. Hayes, "On Modifying Logic Networks to Improve Their Diagnosability", *IEEE Transactions on Computers*, Vol. C-23, No. 1, pp. 56-62, January 1974.

436 L. P. Henckels, "Testability of Logic Circuits Boards," IEEE Intercon, 1974.

437 A. D. Hearn, "Boolean Difference Analysis Program", *IBM Technical Disclosure Bulletin*, Vol. 12, No. 12, p. 2172-73, May 1970.

438 W. Heimerdinger and W. Sterling, "A Program for Generating a Set of Paths for Fault Location," Graph-Mark10, DM006, Illiac-IV-172, 1968.

439 S. Heiss, "Conference Report on the Open Workshop on Fault Detection and Diagnosis in Digital Circuits and Systems," *Computer*, Vol. 4, No. 2, March/April 1971, p. 42.

440 W. C. Hemming, "Functional Simulation Techniques for Design Verification and Fault Insertion in Electronic Digital Systems", Southern Methodist University, Dallas, Texas, Ph.D. Dissertation, August 1971.

441 C. W. Hemming and S. A. Szygenda, "Modular Requirements for Digital Logic Simulation at a Predefined Functional Level," 1972 ACM Conference, p. 380-389.

442 J. M. Hemphill, "The Development of Procedures for the Analysis and Synthesis of a Highly Defined Class of Fault Tolerant Computer System," Southern Methodist University, Dallas, Texas, Ph.D. Dissertation, August 1971.

443 F. C. Hennie, "Fault Detecting Experiments for Sequential Circuits", Proceedings of the 5th Annual Symposium on Switching Theory and Logical Design, p. 95-110, November 1964.

444 F. C. Hennie, *Finite-State Models for Logic Machines*, J. Wiley, 1968.

445 R. I. Hess, "Proposal of an Error-Detecting Test Using Generated Number Sequences," Logicon, Inc., San Pedro, Calif., AIAA Paper No. 69-946, September 1969.

446 R. I. Hess, "An Error-Detecting Test Using Generated Number Sequences", *Journal of Spacecraft and Rockets*, Vol. 7, No. 5, p. 609-612, May 1970.

447 T. N. Hibbard, "Least Upper Bounds on Minimal Terminal State Experiments for Two Classes of Sequential Machines," *Journal of the ACM*, Vol. 8, p. 601-12, October 1961.

448 J. I. Hickman and J. F. Shiner, "Some Results in Computer-Aided Test Generation", IEEE Intercon, 1973.

449 A. N. Higgins, "Error Recovery Through Programming", *FJCC*, 1969, Pt. 1, p. 39-48.

450 L. C. Highby, "Failure Location System for an Airbourne Digital Computer", Third Electronic Industries Association Conference on Maintainability of Electronic Equipment," San Antonio, Texas, December 5-7, 1960.

451 D. Hightower and B. Unger, "A Method of Rapid Testing of Beam Cross-Over Circuits", 9th Annual Design Automation Workshop, Dallas, 26-28 June 1972, p. 144-156.

452 F. J. Hill, E. A. Carter, and B. M. Huey, "A Graph Search Approach to the Determination of Test Sequence for LSI Sequential Circuits", Seventh Asilomar Conference on Circuits, Systems, and Computers, November 27-29, 1973.

453 F. J. Hill and C. S. Meyer, "Interaction with a Simulation for the Determination of Fault Detection Sequences for LSI Circuits", IEEE Resources Roundup, Phoenix Section of the IEEE, April 1969.

454 H. D. Hillman and M. L. Birns, "A Practical Approach to Computer Diagnostics", IEEE Computer Group Repository, R-68-7, August 15, 1967.

455 L. Hillman, "An Automated Dynamic Digital Logic Circuit Test Systems", Computer Design, Vol. 8, August 1969, p. 58-62.

456 Y. Hiyama, Y. Wakabayashi, Y. Shinji, N. Abe and S. Koguchi, "Program-Controlled Memory Test-System", IEEE Transactions on Instrumentation and Measurement, Vol. IM-20, No. 4, p. 242-49, November 1971.

457 D. A. Hodges, "Semiconductor Memories", IEEE Press, 1972.

458 A. J. Hoehn and E. Saltz, "Mathematical Models for the Determination of Efficient Troubleshooting Routes", IRE Transactions on Reliability and Quality Control, Vol. RQC-13, p. 1-14, July 1958.

459 C. E. Holborow, "An Improved Bound on the Length of Checking Experiments for Sequential Machines with Counter Cycles," IEEE Transactions on Computers, Vol. C-21, No. 6, p. 597-98, June 1972.

460 M. E. Homan, "An Approach to the Design of Checking for Combinational Logic", AIEE Conference on the Diagnosis of Failures in Switching Circuits, May 15-16, 1961.

461 L. R. Hoover, "Secondary Techniques for Increasing Fault Coverage of Fault Detection Test Sequences for Asynchronous Sequential Networks", Ph.D. Dissertation, University of Missouri-Rolla, Rolla, Missouri, 1972.

462 L. R. Hoover and J. Tracey, "Fault Coverage and Asynchronous Sequential Networks", Compcon, 1972, p. 239-42.

463 L. R. Hoover and J. Tracey, "Techniques for Increasing Fault Coverage for Asynchronous Sequential Networks", FJCC, 1972, pp. 239-.

464 L. R. Hoover and J. Tracey, "Procedures for Increasing Fault Coverage for Digital Networks", FJCC, 1972, p. 375-84.

465 G. D. Hornbuckle and R. N. Spann, "Diagnosis of Single-Gate Failures in Combinational Circuits", ISSCC, 1969, p. 140-1.

466 G. D. Hornbuckle and R. N. Spann, "Diagnosis of Single-Gate Failures in Combinational Circuits", IEEE Transactions on Computers, Vol. C-18, No. 3, p. 216-20, March 1969.

467 W. A. Hornfeck and C. C. Carroll, "Efficient Generation of Minimum Fault Test Schedules for Combinational Logic Networks", AD-726 383, June 1971.

468 M. S. Horowitz, "Automatic Checkouts of Small Computers", SJCC, 1969, p. 359-65.

469 R. H. House and T. Rado, "Implementation of Logic," *IRE Transactions on Military Electronics*, Vol. MIL-6, p. 297-302, July 1962.

470 D. R. Howarter, "The Selection of Failure Location Tests by Path Sensitizing Techniques", Coordinated Science Laboratory, Report 316, University of Illinois, Urbana, Illinois, August 1966, and AD-640 462.

471 M. Y. Hsiao, F. F. Sellers, and D. K. Chia, "Fundamentals of Boolean Difference for Test Pattern Generation", Proceedings of the Fourth Annual Princeton Conference on Information Science and Systems, March 1970.

472 M. Y. Hsiao and D. K. Chia, "Boolean Difference for Automatic Test Pattern Generation, Part I Combinational Circuits", IBM TR.002149, January 1970, et IEEE Computer Society Repository R.71-92.

473 M. Y. Hsiao and D. K. Chia, "Boolean Difference for Automatic Test Pattern Generation, Open Workshop on Fault Detection and Diagnosis in Digital Circuits and Systems", Lehigh University, Pa., p. 43-44, 7-9 December 1970.

474 M. Y. Hsiao and D. K. Chia, "Boolean Difference for Fault Detection in Asynchronous Sequential Machines", 1971 International Symposium on Fault-Tolerant Computing, p. 9-13, Pasadena, California.

475 M. Y. Hsiao and D. K. Chia, "Boolean Difference for Fault Detection in Asynchronous Sequential Machines", *IEEE Transactions on Computers*, Vol. C-20, No. 11, p. 1356-61, November 1971.

476 M. Y. Hsiao and D. K. Chia, "A Homing Sequence Generation Algorithm for Fault Detection in Asynchronous Sequential Circuits", International Symposium on Fault-Tolerant Computing, p. 137-142, 1972.

477 R. F. Hsiao, "Statistical Approach to Detect a Faulty Partition in a Fault Tolerant Computer System for Fault Diagnosing", IEEE Computer Society Repository, R73-3.

478 E. P. Hsieh, "Optimal Checking Experiments for Sequential Machines", Ph.D. Thesis, Columbia University, 1969.

479 E. P. Hsieh, G. R. Putzolu, and C. J. Tan, "Programmed Test Generators for Sequential Circuits, Open Workshop on Fault Detection and Diagnosis in Digital Circuits and Systems", p. 42, 7-9 December 1970, Lehigh University.

480 E. P. Hsieh, "Checking Experiments for Sequential Machines", *IEEE Transactions on Computers*, Vol. C-20, No. 10, p. 1152-66, October 1971.

481 S. C. Hsieh and R. D. Guyton, "Two Methods for the Design of Checking Experiments", Proc. IEEE Region III, Convention, 1972.

482 S. C. Hsieh, "Two Methods for the Design of Checking Experiments", MS EE. Thesis, Mississippi State University, January 1972.

483 H. H. C. Huang, "Analysis of Random Test Pattern Generation for Combinational Circuits", University of Southern California, Los Angeles, Ph.D. Thesis, January 1971.

484 D. A. Huffman, "Testing for Faults in Cellular Logic Arrays", AD-740 564, January 1972.

485 J. W. Hung, B. Kolman, I. E. Block, and J. G. Hannom, "Investigation and Analysis of Circuit Complexes," AD-600 547, December 1, 1963.

486 J. D. Hutcheson, "An Efficient Method for Generating Fault Tests-for LSI Sequential Circuits", IEEE Computer Society Repository, R.70-94, 24 February 1970.

487 T. Ichikawa and T. Watanabe, "A Method of Finding Diagnostic Test Functions Amenable to Computer Implementation", IEEE Computer Group Repository, R-69-44, 10 December 1968.

488 T. Ichikawa and T. Watanabe, "A Systematic Method of Finding Diagnostic Test Functions", *Electronics and Communications in Japan*, Vol. 52-C, No. 4, p. 165-172, April 1969.

489 V. G. Igolinskiy, "Analytical Methods of Constructing Failure Checkout Tables for Actual Logical Networks", *Engineering Cybernetics*, Vol. 6, No. 11, p. 85-97, 1969.

490 M. Inagaki, "Test and Diagnosis Program Generation Using Microinstruction", *NEC Research and Development*, Vol. 26, No. 7, pp. 35-52, July 1972.

491 J. C. Ippolito and J. P. Richard, "Easily Testable Universal Cellular Network", IEEE Computer Society Repository, R-72-217.

492 J. C. Ippolito and J. P. Richard, "Reseau Cellulaire Universel Facilement Testable", Colloque International Conception et Maintenance des Automatismes Logiques, Toulouse, 27-28 September 1972; p. I-1-26; also N73-12252.

493 R. D. Isenhart, "Optimal Test Sets for the Diagnosis of Multiple Output Combinational Nets", Coordinated Science Laboratory, University of Illinois, R-381, 1968.

494 H. Jacobowitz, "A Logic Oriented Diagnostic Program", *SJCC*, 1967, p. 761-64.

495 H. Jacobowitz, "Automatic Generation of Tests and Diagnostics for Chips, Plug-Ins and Computer Systems", Computers Designer's Conference and Exhibition, Anaheim, 19-21 January 1971.

496 G. C. Jain and H. G. Adshead, "Automatic Exhaustive Testing and Diagnostics of Sequential Logic Networks", 1972 International Symposium on Fault-Tolerant Computing, 19-21 June 1972, p. 73-78.

497 H. Jamet and C. Levy, "Generation Automatique des Tests Pour la Detection et la Localisation des Defauts dans les Ensembles Logiques-Le Programmes Ladislas", Colloque International Conception et Maintenance des Automatismes Logiques, Toulouse, September 27-28, 1972; also N7312204.

498 Jaouen and G. Roux, "Methode de Programmation Pour le Test D'Un Ensemble Logique", CNET, Lannion, Dept. CEI, 28 December, 1966.

499 H. J. Jelinek and H. T. Breen, "Proper Test-Point Allocation", *Electrotechnology*, June 1966, p. 50-3.

500 J. S. Jephson and R. P. MacQuarrie, "A Three Valued Computer Design Verification System," *IEEE International Convention Record*, 1968.

501 J. S. Jephson, R. P. MacQuarrie, and R. E. Vogelsberg, "A Three-Valued Computer Design Verification System", *IBM System Journal*, Vol. 8, No. 3, p. 189-203, 1969.

502 L. R. Johnson, "Diagnosing Hardware via Program-Created Fault Catalog", Preprint 1966, IEEECGR Paper No. 66-49.

503 R. A. Johnson, E. Kletsky, and J. D. Brule, "Diagnosis of Equipment Failures", AD-213 876, April 1959.

504 R. A. Johnson, "An Information Theory Approach to Diagnosis," *IRE Transactions on Reliability and Quality Control*, April 1960, p. 35.

505 R. A. Johnson, "An Information Theory Approach to Diagnosis," 6th National Symposium on Reliability and Quality Control, 1960, p. 102-109.

506 R. A. Johnson and J. D. Brule, "Diagnosis of Equipment Failures," AD-236 188, 15 April 1960.

507 S. M. Johnson, "Optimal Sequential Testing", *The Rand Corporation Research Memo* 1652, 1956.

508 W. R. Johnson,"Proving Out Large PC Boards, Which System is Best for You," *Electronics*, March 15, 1971, p. 68-71.

509 D. M. Jones, H. G. Shah, and B. D. Carroll, "On the Generation of Tests for Multiple Faults", *IEEE Transactions on Computers*, 1973.

510 Jones, D. M. , ph.D. Dissertation, Auburn Univ., 1973.

511 D. M. Jones, H. G. Shah, and B. D. Carroll, "On the Generation of Tests for Multiple Faults", IEEE Computer Society, R73-26.

512 E. R. Jones and C. H. Mays, "Automatic Test Generation Methods for Large Scale Integrated Logic", *IEEE Journal of Solid State Circuits*, Vol. SC-2, No. 4, p. 221-26, December 1967.

513 E. R. Jones, "Automatic Test Synthesis", *Wescon Technical Papers*, p. 23-26, August 1969.

514 J. R. Jump and D. R. Fritsche, "Microprogrammed Arrays", *IEEE Transactions on Computers*, Vol. C-21, No. 9, p. 974-84, September 1972.

515 V. V. Kabrinskii, P. P. Parkhomenko, and E. S. Sogomonyan, "Technical Diagnostics of Monitored Plants", Energiya, 1967.

516 K. Kajitani, Y. Tezuka, and Y. Kasahara, "Diagnosis of Multiple Faults in Combinational Circuits", *Electronics and Communications in Japan*, Vol. 52-C, p. 123-131, April 1969.

517 S. Kamal, "Intermittent Faults-A Model and a Detection Procedure," Computer Science Conference, Columbus, Ohio, February 20-22, 1973.

518 S. Kamal and C. W. Page, "Intermittent Faults—A Model and a Detection Procedure," IEEE Computer Society Repository, R73-249.

519 S. Kamal and B. Weinberg, "Diagnostic Test Selection," IEEE Computer Society Repository, R73-84.

520 Y. Kambayashi and S. Yajima, "Fault Detecting Experiments for Finite-Memory Sequential Machines and Their Applications," Papers of Technical Group on Automation and Information Theory, IECE Japan, AIT 68, 15 June 1968.

521 Y. Kambayashi and S. Yajima, "Controllability of Sequential Machines", *Information and Control*, Vol. 21, pp. 306-328, November 1972.

522 K. Kanda, "Computerized Product System Fault Trees."

523 J. M. Kaplan, "Software Diagnosis of Memory Failures," *Honeywell Computer Journal*, January 1970.

524 A. A. Kaposi, "Logic Testing by Simulation," IEE Publication 51 on CAD 1969, p. 31-40.

525 A. A. Kaposi and D. R. Holmes, "Transient Testing of Logic Networks," *Computer Aided Design*, Vol. 3, No. 2, p. 23-28, Winter 1971.

526 A. A. Kaposi, "Logic Design Testing", Ph.D. Thesis, 1971.

527 A. A. Kaposi, "On the Testability of Digital Hardware," *Computer Aided Design*, July 1972, p. 169-171.

528 M. F. Karavai, "Diagnosis of Single Failures of the Type of Logical Constant in

Combinational Circuits", *Automation and Remote Control*, Vol. 32, No. 9, p. 1501-7, September 1973.

529 M. K. Karavai, "Construction of Tests for Finding Multiple Faults in Combinational Devices in Arbitrary Bases," *Automation and Remote Control*, Vol. 34, No. 4, pp. 656-67, April 1973.

530 V. V. Karibiskii, P. P. Parkhomenko and Y. S. Sogomonyan, "Engineering Diagnostic of Composite Devices", Coll. Abstract and Structural Theory of Relay Devices, Nauka 1961.

531 V. V. Karabiskii, "Analysis of Systems for the Purpose of Monitoring and Diagnosing Failures", *Automation and Remote Control*, 1964.

532 V. V. Karibiskii, "The Analysis of Systems for Checking Operation and Diagnosing Faults", *Automation and Remote Control*, Vol. 26, No. 2, p. 305-310, 1965.

533 V. V. Karibiskii, P. P. Parkhomenko, and E. S. Sogomonyan, "The Construction of Monitoring and Diagnostic Tests for Discrete and Continuous Units," Third All-Union Conference on Automatic Control, September 20-26, 1965, Paper No. 87.

534 V. V. Karibiskii, P. P. Parkhomenko, and E. S. Sogomonyan, "Some Problems of Checking the Performance and Locating Failures in Finite Automata," *Soviet Physics Doklady*, Vol. 10, No. 3, p. 182-194, 1965.

535 V. V. Karibiskii, P. P. Parkhomenko, and E. S. Sogomonyan, "Diagnostics for Combinational Circuits", Third Congress of the International Federation of Automatic Control, June 20-25, 1966, London, Paper 35D.

536 V. V. Karibiskii, P. P. Parkhomenko and E. S. Sogomonyan, "Engineering Diagnostics of Combinational Devices," In Abstraktnaya I Strukturnaya Teoriya Releinykh Ustroistv, p. 189-224(M. A. Gavrilov Ed.) (IZDAT. Nauka, Moscow 1966).

537 V. V. Karibiskii, "Construction of an Input Sequence That Detects a Given Failure of a Discrete Device," *Automation and Remote Control*, Vol. 33, No. 5, pp. 843-51, May 1972.

538 V. V. Karibiskii, "Setting of Memory Elements in Given States and Failure Detection in a Discrete Device," Proceedings of the First All-Union Conference on Technical Diagnostic, Nauka, 1972.

539 R. M. Karp and J. P. Roth, "Derivation of a Diagnosis Procedure," Research Memorandum ST-101, IBM Corporation, 5 January 1960.

540 R. M. Karp and E. G. Manning, "Automatic Compilation of a Digital Automation Language for Digital Fault Simulation," 1969.

541 T. Kasami, A. Hashimoto, and H. Ozaki, "A Study of the Methods for Detecting Fault Elements of Combinational Logical Networks", *Studies of Electronic Computers*, 2, 1962.

542 M. Kato, Y. Koga, and K. Naemura, "Diagnostic Test Patterns and Sequences for Illiac IV Pe," Illiac IV 180, 1968.

543 W. H. Kautz, "Automatic Fault Detection in Combinational Switching Networks", Proceedings Symposium of Switching Theory and Logic Design, p. 195-214 and AD-267 005, April 1961.

544 W. H. Kautz, "Diagnosis and Testing of Cellular Arrays, Properties of Cellular Arrays for Logic and Storage", SRI Project 5876, Scientific Report 3, p. 119-145, July 1967.

545 W. H. Kautz, "Fault Diagnosis in Combinational Digital Circuits", Digest First Annual IEEE Computer Conference, p. 2-5, September 1967.

546 W. H. Kautz, "Testing for Faults in Combinational Cellular Logic Arrays", Proceedings of the 8th Annual Symposium on Switching and Automata Theory, p. 161-74, October 1967.

547 W. H. Kautz, "Fault Testing and Diagnosis in Combinational Digital Circuits", IEEE Transactions on Computers, Vol. EC-17, No. 4, p. 352-66, April 1968.

548 W. H. Kautz, "Some Classes of Minimally Testable Combinational Networks", Open Workshop on Fault Detection and Diagnosis in Digital Circuits and Systems, p. 9, 7-9 December 1970, Lehigh University.

549 W. H. Kautz, "Testing for Faults in Wiring Networks", IEEE Computer Society Repository, R-72-212, and Stanford Research Institute.

550 T. Kawata, S. Murakami, K. Kinoshita, and H. Ozaki, "A Consideration on Fault Detection of Feedback Shift-Register Circuits", Joint Convention of Four Institutes for Electrical Engineering, Japan, No. 2687, 1968.

551 T. Kawata, K. Kinoshita, and H. Ozaki, "A Method for Fault Detection in Feedback Shirt-Register Circuits", Electronics and Communications in Japan, Vol. 52C, No. 7, p. 85-92, July 1969.

552 V. I. Kaznacheev, "The Design of Tests for Finite Automata with the Aid of a Language of Control Expressions", Problemy Sinteza Tsifrovykh Avtomatov, p. 145-158 (IZDAT. Nauka, Moscow 1967).

553 V. I. Kaznacheev, "A Model of a Digital Automaton with Malfunctions", Izvestiya A. N. SSSR, Tekhnicheskaya Kibernetica, No. 6, p. 55-62, 1968.

554 L. Kedson and A. M. Stoughton, "Memory Testing is a Task That Comes in Layers", Electronics, 1 September 1969, p. 88-96.

555 W. L. Keiner, "Functional Testing—A User Looks at Logic Simulation", 10th Design Automation Workshop, 1973.

556 J. Kella, "Sequential Machine Identification", IEEE Transactions on Computers. Vol. C-20, No. 3, p. 332-338, March 1971.

557 R. M. Keller and D. F. Wann, "Analysis of Implementation Errors in Digital Computing Systems", Technical Report No. 6, March 1968, Computer Systems Laboratory, Washington University, St. Louis, Missouri, and AD-669 812.

558 B. De Kersauzon and A. Royer, "Outils D'Etude Par Simulation des Propagations de Defauts dans les Materiels Logiques," Congres de L'Afcet 1970.

559 V. F. Khalchev, "Construction of a Checking Test for Combinational Circuits," Automation and Remote Control, October 1970.

560 C. R. Kime, "A Failure Detection Method of Sequential Circuits, Department of Electrical Engineering," Technical Report 66-13, January 1966, University of Iowa.

561 C. R. Kime, "An Organization for Checking Experiments on Sequential Circuits", IEEE Transactions on Electronic Computers, Vol. EC-15, p. 113-115, February 1966.

562 C. R. Kime, "Checking Experiment Generation for Synchronous Sequential Circuits", IEEE Computer Group Repository R-67-65, 27 December 1966.

563 C. R. Kime and J. Ellenbecker, "The Generalized Fault Table and Its Use in Diagnostic Test Selection," NEC 1969, p. 663-667.

564 C. R. Kime, "A Diagnosability Analysis Model for Digital Systems", Proceedings of the 2nd Hawaii International Conference on System Sciences, Western Periodical 1969, p. 881-884.

565 C. R. Kime, "An Analysis Model for Digital System Diagnosis," *IEEE Transactions* on Computers, Vol. C-19, No. 11, p. 1063-73, November 1970.

566 C. R. Kime, "Comment on 'Derivation of Minimal Complete Sets of Test-Input Sequences Using Boolean Differences' ", IEEE Computer Society Repository, R-71-172.

567 K. Kinoshita, "Some Considerations on the Fault Diagnosis of Sequential Circuits", *Electronics and Communications in Japan*, Vol. 46, No. 9, p. 20-28, September 1963.

568 K. Kinoshita, H. Ozaki, and S. Murakami, "Computer Experiments for Finding Fault-Detecting Sequences of Sequential Circuits", *Electronics and Communications in Japan*, Vol. 50, p. 191-201, 1967.

569 N. V. Kinsht, "A Procedure for Malfunction Search", Avtometryia No. 3, p. 34-38, 1965 (Abstract in USSR Scientific Abstracts, C. C. A. T., No. 11, p. 134).

570 N. V. Kinsht, "Criteria for the Optimization of the Fault Detection Process," Izvestiya Sibirskovo Otdeleniya A. N. SSSR, *Seriya Tekhnicheskikh Nauk*, Fol. 3, No. 10, 1965.

571 N. V. Kinsht, "Some Problems in Optimizing the Procedure of Restoring Failure-Free Operation of Systems Using Testing and Substitution of Elements", *Avtometryia*, No. 4, p. 118-126, 1966.

572 N. V. Kinsht, "Optimal Procedures for Fault Detection in a Class of Models of Diagnostic Systems," Conference on Automatic Testing and Methods of Electrical Measurement, IZDAT, Sibirskoe Otdelnie A. N. SSSR, 1967.

573 A. R. Klayton, *et al*, "The Detection and Diagnosis of Memory System Faults", 8th Design Automation Workshop, 1971.

574 A. R. Klayton and A. K. Susskind, "Multiple Fault-Detection Tests for Loop-Free Logic Networks", 1971 International IEEE Computer Conference, p. 77-78.

575 A. R. Klayton, "An Algebraic Approach to Fault Detection", Open Workshop in Fault Detection and Diagnosis in Digital Systems, Lehigh University, 6-8 December 1971.

576 E. J. Kletsky, "An Application of the Information Theory Approach to Failure Diagnosis", *IRE Transactions on Reliability and Quality Control*, p. 29-39, December 1960.

577 E. J. Kletsky, "Diagnosis of Equipment Failures", Pt. 2, Rome Air Development Center, Technical Report 60-67-B, April 1960, and AD-236 189.

578 K. L. Kodandapani and S. Swamy, "Diagnosis of Single Cell Failures in Cutpoint Cellular Arrays", IEEE Computer Society Repository, R-72-235.

579 K. L. Kodandapani, "Fault Diagnosis in AND-EX-OR Circuits," IEEE Computer Society Repository, R73-117.

580 K. L. Kodandapani, "Comments on Easily Testable Realizations for Logic Functions", IEEE Computer Society Repository, R73-320.

581 B. V. Koen, "Methodes Nouvelles Pour L'Evaluation de la Fiabilite-Reconnaissance des Formes-I-Quelques Idees Nouvelles Pour L'Evaluation de la Fiabilite, II Application de la Reconnaissance des Formes aux Calcul de la Fiabilite des

Systemes", *Cen Saclay Departement Surete Nucleaire*, CEA-R-4368, 1972, pp. 1-112.

582 Y. Koga, "Methods for Generating Diagnostic Sequences," University of Illinois, Department of computer Science, Illiac Document No. 100, 1967.

583 Y. Koga, "A Checking Method of Wiring," IEEE Computer Group Repository, R-70-77, 26 January 1970.

584 Y. Koga, "A Checking Method of Wiring", 7th Annual Design Automation Workshop, 1970.

585 Y. Koga, T. C. Chen, and K. Naumura, "A Method of Test Generation for Fault Location in Combinational Logic", *FJCC*, 1970, p. 69-73.

586 Y. Koga and F. Hirata, "Fault-Locating Test Generation for Combinational Logic Networks", 1972 International Symposium on Fault-Tolerant Computing, 19-21 June 1972, p. 131-36.

587 I. V. Kogan, "Monitoring the Operation of Logical Devices," Doklady na Mezhdunarodnom Simposiume Po Teorii Releinykh Ustroistv I Konechnykh Avtomatov (Moscow 1962), Transl. in Relay Systems and Finite Automata, p. 116-137, Burroughs Corp. Paoli, Pennsylvania, 1964.

588 I. V. Kogan, "The Construction of Tests for Contact Systems", Proceedings of the Seminar on Problems in the Theory of Mathematical Digital Computers, Scientific Council on Cybernetics, A. N. USSR, Kiev, 1963.

589 I. V. Kogan, "Tests for Nonrepetitious Contact Networks Problem", *Kibernetiki*, No. 12, p. 39-44, 1064.

590 I. V. Kogan, "The Problem of the Diagnostic Testing of Digital Computers", *Tsifrovaya Vychislitelnaya Tekhnika I Programmirovanie*, No. 2, p. 66-69, 1967, Transl. in N70-14631, 1970.

591 I. Kohavi and Z. Kohavi, "New Techniques for the Design of Fault Detection Experiments for Sequential Machines", Proceedings of the First Hawaii International Conference of System Sciences, p. 810, January 1968.

592 I. Kohavi and Z. Kohavi, "Variable Length Distinguishing Sequences and Their Application to the Design of Fault Detection Experiments", *IEEE Transactions on Computers*, Vol. EC-17, No. 4, p. 792-95, August 1968.

593 I. Kohavi, "Fault Diagnosis of Logical Circuits", 10th Annual Symposium on Switching and Automata Theory, p. 166-73, 1969.

594 I. Kohavi, "Fault Detection Experiments for Logical Networks", Ph.D. Dissertation, Polytechnic Institute of Brooklyn, 1970.

595 I. Kohavi, "Detection of Faults in Combinational Logic Networks", IEEE Computer Group Repository, R-70-110.

596 I. Kohavi and Z. Kohavi, "Detection of Multiple Faults in Combinational Logic Networks", *IEEE Transactions on Computers*, Vol. C-21, No. 6, p. 556-568, June 1972.

597 Z. Kohavi and P. Lavallee, "Design of Diagnosable Sequential Machines", *IEEE Transactions on Electronic Computers*, Vol. EC-16, No. 4, p. 473-85, August 1967.

598 Z. Kohavi and P. Lavallee, "Design of Diagnosable Sequential Machines", *SJCC*, 1967, p. 713-18.

599 Z. Kohavi, *Switching and Finite Automata Theory*, McGraw Hill, New York, 1970.

600 Z. Kohavi and D. A. Spires, "Detection of Failures in Combinational Digital Circuits", *PIEE*, Vol. 118, No. 5, p. 643-8, May 1971.

601 Z. Kohavi, "Theory of Machines and Computations,"International Symposium on the Theory of Machines and Computations, Technion, Haifa, August 1971, Academic Press, New York, 1971.

602 Z. Kohavi and D. A. Spires, "Designing Sets of Fault-Detection Tests for Combinational Logic Circuits", *IEEE Transactions on Computer*, Vol. C-20, No. 12, p. 1463-79, December 1971.

603 Z. Kohavi, J. A. Rivierre, and I. Kohavi, "Machine Distinguishing Experiments", *The Computer Journal*, Vol. 16, No. 2, pp. 141-147, May 1973.

604 K. Kojima, C. Tanaka, and H. Kanada, "Generation and Reduction of Diagnostic Data", IEEE Computer Society Repository, R-72-181.

605 N. R. Kornfield, "Boolean Matrices and Their Application to Fault Reduction in Switching Networks", Ph.D. Thesis, University of Pennsylvania, 1964.

606 L. A. Korytnaya, "Automatic Detection of Malfunctions in Electronic Digital Machines", *Cybernetics Techniqeus*, Kiev, 1965, p. 90-101, et Rept. FTD-HT23-599-67., Wright Patterson AFB, Ohio, and AD-689 260, 27 July 1967.

607 E. S. Kospanov, "An Algorithm for Constructing Very Simple Tests", *Diskretnyi Analiz*, No. 8, p. 43-48, 1966.

608 N. Kouvaras and D. Lagoyannis, "Mathematical Analysis of Failures in Sequential Circuits", *Electronic Engineering*, Vol. 40, No. 481, p. 156-8, March 1968.

609 D. Kramer, "Test Criteria for Logic Circuits", *Messen Steuren Regeln*, Vol. 15, No. 8, p. 288-93, August 1972.

610 T. Kranton and A. Libenson, "A Pattern Recognition Approach to Fault Isolation", IEEE Transactions on Aerospace Support Conference Procedure, AS-1, No. 2, p. 1320-1326, August 1963.

611 D. A. Kristinkoy, "A Method for Assessing Malfunctions in Complex Mechanical Systems", *Kibernetika I Diagnostika*, (Sbornik) No. 1, p. 54-62, 1966.

612 T. A. Kriz, "A Path Sensitizing Algorithm for Diagnosis of Binary Sequential Logic", IEEE Computer Group Repository, 1969.

613 T. A. Kriz, "Machine Identification Concepts of Path Sensitizing Fault Diagnosis", 10th Annual Symposium on Switching and Automata Theory, p. 174-81, October 1969.

614 T. A. Kriz, "A Path Sensitizing Algorithm for Diagnosis of Binary Sequential Logic", IEEE International Computer Conference, p. 250-59, June 1970.

615 T. A. Kriz, "Use of Path and Tree Diagnostic Techniques for Digital Fault Isolation," Open Workshop on Fault Detection and Diagnosis in Digital Systems, Lehigh University, 6-8 December 1971.

616 S. P. Krosner, "Test Data Generation Tutorial", Proceedings of the ACM Annual Conference, August 1972, p. 1146-47.

617 J. B. Kruskal and R. E. Hart, "A Geometric Interpretation of Diagnostic Data from a Digital Machine", *B.S.T.J.*, Vol. 45, No. 8, p. 1299-1338, October 1966.

618 H. Kubo, "A Procedure for Generating Test Sequences to Detect Sequential Failures", *NEC Research and Development*, No. 12, p. 69-78, October 1968.

619 S. N. Kukreja and I-Ngo Chen, "Combinational and Sequential Cellular Structures",

IEEE Transactions on Computers, Vol. C-22, No. 9, pp. 813-23, September 1973.

620 P. I. Kuznetsov and L. A. Pchelintsev, "A Problem of Fault Searching", *Automation and Remote Control*, No. 3, 1969, p. 137-140.

621 P. I. Kuznetsov and L. A. Pchelintsev, "The Application of Some Mathematical Methods in Medical Diagnostics", *Math. Biosci.*, Vol. 5, Nov. 1969, pp. 365-377.

622 F. Kvamme, "Standard Read-Only Memories Simplify Complex Logic Design", *Electronics*, 5 January 1970, p. 88-95.

623 I. Lajovic, "Diagnose—A Language for Description of Diagnostic Problems", *Automatika* (Yugoslavia), Vol. 12, No. 6, pp. 374-80, 1971.

624 S. E. Lamacchia, "Diagnosis in Automatic Checkout", *IRE Transactions on Military Electronics*, Vol. MIL-6, No. 3, p. 302-309, July 1962.

625 H. R. Lambert, "The Characteristics of Faults in MOS Arrays", Workshop on Hardware Faults and the Programmer, Palm Springs, 28-29 June 1968.

626 H. R. Lambert, "Characteristics of Faults in MOS Arrays", *SJCC*, 1969, p. 403-410.

627 B. Lampe, "Automatic Fault Diagnosis in an Electronic Switching System, Design of the Modules", International Conference on Communications, June 11-13, 1973.

628 R. E. Lander, "Computer Testing by Control Waveform Simulation", *Computer Technology*, IEE Conference, Publication No. 32, p. 9-19, 1967.

629 R. W. Landgraff and S. S. Yau, "Design of Diagnosable Iterative Arrays", *IEEE Transactions on Computers*, Vol. C-20, No. 8, p. 867-77, August 1971.

630 G. G. Langdon and C. K. Tang, "Concurrent Error Detection for Group Lookahead Binary Adders", IEEE Workshop on Reliability and Maintainability, Ozarks, Missouri, 1969.

631 R. W. Larsen, "Error-Tolerant Sequential Circuit Using Error-Correcting Codes", AD-710 210, June 1970.

632 B. Lattin, "Testing Dynamic Rams", *Wescon Technical Papers*, 1971, p. 304-308.

633 L. J. Lauler, R. B. Whiteley, and D. E. Sailor, "Automatic Check-Out Equipment Featuring Test Programs for Diagnosis Checking", *IRE 1959 National Convention Record*, Vol. 7, Part 4, p. 218-222.

634 B. A. Laws and C. K. Rushforth, "A Cellular-Array Multiplier for GF(2*M)", Montana State University, Bozeman, IEEE Computer Group Repository, R-70-178.

635 J. Lederberg, "System for Computer Construction Enumeration and Notation of Organic Molecules as Tree Structures and Cyclic Graphs", Interim Report to the NASA, Grant No. G.81-60, Part 2, December 1965.

636 H. P. Lee and E. S. Davidson, "Redundancy Testing in Combinational Networks", IEEE Computer Society Repository, R73-212.

637 A. Y. Lee, "An Algorithm for Testing K—Controllability and Generating the Necessary Input Sequences of Nonlinear Sequential Networks", NEC 1972, pp. 100-103.

638 P. M. Lee, "Practical Aspects of LSI Testing", IEEE International Convention Record, 1971, p. 556-57.

639 S. C. Lee and E. T. Lee, "Fault Detection of Neural Networks Automata Made Up of Parts", New York University, Bronx, IEEE Computer Group Repository, R-70-134.

640 S. C. Lee and E. T. Lee, "Fault Diagnosis of Neural Networks", Proceedings of the Third Hawaii Conference, January 1971, p. 739-742.

641 N. Legrand, "Programme de Detection et de Localisation des Defauts dans les Circuits Logiques Combinatoires et Sequentiels (Delod)", Note Technique Thomson-CSF, No. LCR-DR5-1809, June 23, 1972.

642 M. Lehman, R. Eshed, and Z. Netter, "Checking of Computer Logic by Simulation on Computers", *Computer Journal*, Vol. 6, No. 2, p. 154-62, July 1963.

643 J. Lellouch and A. Alperovitch, "Une Note Sur L'Application de L'Algorithme de Ford a un Probleme de Diagnostic Medical", *Revue Francaise D'Informatique et de Recherche Operationnelle*, Vol. 5, No. V-1, pp. 57-59, January 1971.

644 V. I. Levin, "Calculation of the Probability of Correct Operation of an Automation with Possible Failures, in Problems of Synthesis of Finite Automata, V. I. Levin, Ed. p. 109-116, Riga, USSR, Izdatel'Stvo, *Zinatne*, 1972.

645 C. Levy, "Detection et Localisation des Avaries dans les Ensembles Logiques", Congres Afcet 1970, and DGRST Protocole, No. 69-01-609.

646 H. M. Levy and D. W. Bray, "Fault Detection Methods in Combinational Digital Logic Networks", AD-760543, January 1973.

647 D. Lewis, "Hazard Detection by a Quinary Simulation of Logic Devices with Bounded Propagation Delays", 9th Annual Design Automation Workshop, Dallas, 26-28 June 1972, p. 157-164.

648 R. S. Lewis, "An Approach to Test Pattern Generation for Synchronous Sequential Circuits", Ph.D. Thesis, Southern Methodist University, 22 December 1967.

649 F. Liguori, "State-of-the Art Computer-Aided Test Design Tools", *Wescon*, 1973.

650 W. R. Lile and R. J. Hollingsworth, "Design Processing and Testing of LSI Arrays for Space Station", N73-13732, October 1972.

651 B. M. Lin and J. P. Robinson, "A Complete Test Set for Quasi Prime Implicant Networks", AD-730 780, August 1971.

652 B. M. Lin and J. P. Robinson, "Testability of Quasi Prime Implicant Networks", AD-740 088, March 1972.

653 B. M. Lin and J. P. Robinson, "Testability of Quasi Prime Implicant Networks", Computer Science Conference, Columbus, Ohio, February 20-22, 1973.

654 M. Ludlow, "The Diagnosis of Intermittent Faults on Logic Circuits", Plessey, October 1973.

655 A. A. Lyapunov, "Logical Schemes of Programs", *Problems of Cybernetics*, Vol. 1, p. 48-81, Pergamon Press.

656 Yu. V. Lyubatov, "Optimization of a Method of Detecting Faults in a Complex System", Trudy Seminara Po Tochnom Machinostroenii i Proborostroenii, Institut Mashinnovedeniya, No. 17, p. 89-97, 1963.

657 Yu. V. Lyubatov, "Optimal Procedures for the Localization of Breakdowns in a Modularized Radioelectronic System, *Engineering Cybernetics*, No. 4, p. 14-21, 1964.

658 Yu. V. Lyubatov, "Some Optimal Methods for Searching for Malfunctions in Engineering Systems", In Nadezhnost Slozhnykh Tekhnicheskikh Sistem (N. G. Davidova Ed.) p. 157-75 (Izdat-Sovetskoe Radio, Moscow, 1966).

659 Yu. V. Lyubatov and M. H. Gaimalov, "A Bound on the Diagnostic Efficiency of Tests", *Engineering Cybernetics*, No. 5, 1968, p. 102-106.

660 K. Maling and E. L. Allen, "A Computer Organization and Programming System for Automated Maintenance", *IEEE Transactions on Electronic Computers*, Vol. EC-12, No. 6, p. 887-895, December 1963.

661 K. Maling, "Classes of Component Failures and a System Design to Facilitate Their Location", TR.00-1029 Rev. IBM Data System Division, Dev. Lab. Poughkeepsie, New York, 28 June 1963, Rev., 10 March 1964.

662 Yu. V. Malyshenko, "Approach to Malfunction Diagnostics in Combination Circuits", *Automation and Remote Control*, Vol. 34, No. 4, pp. 668-70, April 1973.

663 Mandelbaum, D., "Unequal Error Protection Codes Derived from Difference Sets", IEEE Trans. on Info. Thy., Vol. IT-18, pp. 696-697, 1972.

664 D. Mandelbaum, "A Measure of Efficiency of Diagnostic Tests upon Sequential Logic", *IEEE Transactions on Electronic Computers*, Vol. EC-13, p. 630, October 1964.

665 D. Mandelbaum, "On Error Control in Sequential Machines", *IEEE Transactions on Computers*, Vol. C-21, No. 5, p. 492-95, May 1972.

666 E.G. Manning,"On Computer Self-Diagnosis—Part I— Experimental Study of a Processor, Part-II—Generalizations and Design Principles", *IEEE Transactions on Electronic Computers*, Vol. EC-15, p. 873-90, December 1966.

667 E. G. Manning and H. Y. Chang, "A Comparison of Fault Simulation Methods for Digital Systems", Digest of the First Annual IEEE Computer Conference, p. 10-13, September 1967.

668 E. G. Manning and H. Y. Chang, "Functional Techniques for Efficient Digital Fault Simulation", *IEEE International Convention Digest*, 1968, p. 194.

669 P. N. Marinos, "Boolean Differences as a Means for Deriving Complete Sets of Input Test Sequences for Fault Diagnosis in Switching Circuits", Bell Labs Technical Memo No. MM-69-6313-1, August 1969.

670 P. N. Marinos, "A Method of Deriving Minimal Complete Set of Test—Input Sequences Using Boolean Differences", IEEE International Computer Conference, p. 240-6, June 1970.

671 P. N. Marinos, "Partial Boolean Differences and Their Application to Fault Detection", Fourth Asilomar Conference on Circuits and Systems, Pacific Grove, November 1970.

672 P. N. Marinos, "Derivation of Minimal Complete Sets of Test-Input Sequences Using Boolean Differences", *IEEE Transactions on Computers*, Vol. C-20, No. 1, p. 25-32, January 1971.

673 P. N. Marinos, "The Partial Boolean Differences Algorithm—Implementation Strategies", IEEE Computer Group Repository, R-71-90.

674 P. N. Marinos, E. N. Page and M. G. Thomasson, "Time Independent Boolean Differences", IEEE Computer Group Repository, R-71-120.

675 P. N. Marinos, "Fault Diagnosis in Asynchronous Switching Networks Using Partial Boolean Difference", Proceedings of the 9th Annual Symposium IEEE Region III Convention, April 1971, p. 61-67.

676 P. N. Marinos and E. W. Page, "Derivation of Test-Input Sequences for Detecting Short Circuits in Logical Networks", Proceedings of the Third Annual Southeastern Symposium on System Theory, Vol. 1, p. E2-0/E2-8, 1971.

677 P. N. Marinos, "Fault Diagnosis in Digital Systems—An Overview", 1971 IEEE International Computer Conference, p. 71.

212 RATIONAL FAULT ANALYSIS

678 P. N. Marinos, E. M. Page, and M. G. Thomasson, "Fault Detection in Sequential Networks Using Time Dependent Boolean Differences", 1971, International IEEE Computer Society Conference, p. 79-80.

679 R. A. Marlett and C. Fok, "Failure Mode Analysis and Pattern Generation", 1971 International Convention Digest, p. 554-55.

680 R. A. Marlett, "A Test Generation System to Facilitate Diagnostics and Trouble-Shooting," Open Workshop on Fault Detection and Diagnosis in Digital Systems, Lehigh University, 6-8 December 1971.

681 R. A. Marlett, "Automated Digital Test Program Generation for Sequential and Combinational Logic", Proceedings National Nepcon, 1972, p. 138-142.

682 N. A. Martellato and R. H. Wayne, "On Test Conditions for Combinational Logic Circuits", Bell Telephone Laboratories, Murray Hill, New Jersey, 1963.

683 R. L. Martin, "The Design of Diagnosable Sequential Machines", Proceedings of the Hawaii International Conference on System Sciences, p. 619, January 1968.

684 R. L. Martin, "Studies in Feedback Shift Register—Synthesis of Sequential Machines", MIT Press, 1969.

685 V. Martin, "Obtention D'Un Ensemble de Tests Minimaux," Centre D'Informatique de Toulouse, June 1970.

686 L. L. Mate, H. Y. H. Chuang, and S. Das, "A New Approach to Logic Hazard Detection and Elimination", Computer Science Conference, Columbus, Ohio, February 20-22, 1973.

687 R. J. Matteis, "Relevance of Statistical Decision Theory to Fault Isolation", 27th National Meeting of Operations Research, 1965.

688 T. Matuszewski and D. L. Plato, "Computer Generated Testing Programs for Wired Equipment", *The Western Electric Engineer*, Vol. 9, p. 16-23, July 1965.

689 W. Mayeda, "Distinguishability Criteria and Diagnosis in Computers, II," Twelfth Midwest Symposium on Circuit Theory, p. 21-22, April 1969, and AD-690 127.

690 W. Mayeda and C. V. Ramamoorthy, "Distinguishability Criteria in Oriented Graphs and Their Application to Computer Diagnosis 1", *IEEE Transactions on Circuit Theory*, Vol. CT-16, No. 4, p. 448-54, November 1969, and AD-683 739.

691 W. Mayeda and C. V. Ramamoorthy, "Computer System Diagnosis Using Test Gates Method", International Symposium on Fault-Tolerant Computing, March 1971, p. 122-125.

692 W. Mayeda, "Graph Theoretic Computer Diagnosis", *NEC 1971*, pp. 347-350.

693 E. H. Mayer, "An Approach to a Standard Topology for Digital Hybrid Integrated Circuits", Proceeding on Electronic Components, p. 358-44, 1971.

694 V. Mayper, "Problems and Pitfalls in Automatic Test Computer Programming", No. Sete 210-86, NYU Press, May 1965.

695 R. B. McClee, "Some Aids to the Detection of Hazards in Combinational Swtiching Circuits", *IEEE Transactions on Computers*, Vol. C-1, No. 6, p. 561-565, June 1969.

696 R. M. McClure, "Test-Programs for Logic Boards", *Intercon 1972*, p. 514-15.

697 R. M. McClure, "Fault Simulation of Digital Logic Utilizing a Small Host Machine", Proceedings of the 9th ACM-IEEE Design Automation Workshop, June 1972, p. 104-110.

698 E. J. McCluskey, "Determination of Redundancies in a Set of Patterns", *IRE*

Transactions on Information Theory, Vol. IT-13, pp. 167-168, June 1957.

699 E. J. McCluskey, "Transients in Combinational Logic Circuits", In *Redundancy Techniques for Computing Systems*, Spartan Books, 1962, pp. 9-46.

700 E. J. McCluskey, "Test and Diagnosis Procedures for Digital Networks", *Computer*, January-February 1971, p. 17-20.

701 E. J. McCluskey and F. W. Clegg, "Fault Equivalence in Combinational Networks", *IEEE Transactions on Computers*, Vol. 6, C-20, No. 11, p. 1286-93, November 1971.

702 J. F. McKewitt, "Parity Fault Detection in Semiconductor Memories", *Computer Design*, July 1972, p. 67-73.

703 L. P. McNamee, "High Order Language Requirements for Electronic Testing", *NEREM Record, 1969*, p. 222-23.

704 R. E. Meagher and J. P. Nash, "The Ordvac", *Review of Electronic Digital Computers*, 1952, p. 37-43.

705 R. J. Meckert and A. M. Stoughton, "Computer-Controlled Testing of Ferrite Memory", *IEEE Transactions on Magnetics*, Vol. MAG-6, No., p. 38-39, March 1970.

706 M. A. Mehta, "A Method for Optimizing the Diagnostic Resolution of Input-Output Faults in Digital Circuits", Ph.D. Thesis, Illinois Institute of Technology, 1971.

707 M. A. Mehta, H. P. Messinger, and W. B. Smith, "Functions for Improving Diagnostic Resolution in an LSI Environment", *SICC 1972*, p. 1079-1091.

708 Mehta, M. A., and J. G. Valassis, "Simulated Failure Analysis of Integrated Circuits", Auto. Elec. Tech. Jour., Vol. 11, pp. 306-316, 1969.

709 M. A. Mehta and W. B. Smith, "On Further Optimization of AR Functions", IEEE Computer Society Repository, R73-233.

710 K.C.Y. Mei, "Fault Dominance in Combinational Circuits", Technical Note No. 2, Digital Systems Laboratory, Stanford Electronics Labs, Stanford University, California, August 1970.

711 K. C. Y. Mei, "Bridging and Stuck-At Faults", 1973 International Symposium on Fault-Tolerant Computing, Palo Alto, California, June 20-22, 1973.

712 K. C. Y. Mei, "Fault Dominance in Combinational Circuits", Workshop on Diagnosis and Reliable Design of Digital Systems, Pasadena, December 5-7, 1973.

713 K. C. Y. Mei, "Bridging and Stuck-At Faults", IEEE Computer Society Repository, R73-257.

714 K. Menger, "Checkups for Combinational Gates," *IEEE Transactions on Aerospace Support Conference Procedure*, Vol. AS-1, No. 2, p. 954-60, August 1963.

715 P. R. Menon, "A Simulation Program for Logic Networks", Bell Telephone Labs, Internal Memo No. MM-65-1271-3, 1965.

716 P. R. Menon, "Design of Generalized Double Rank and Multiple Rank Sequential Circuits", *Information and Control*, Vol. 15, p. 436-51, November 1969.

717 P. R. Menon and A. D. Friedman, "Fault Detection in Iterative Logic Arrays", *IEEE Transactions on Computers*, Vol. C-20, No. 5, p. 524-535, May 1971.

718 G. Metze, "Functional Diagnosability and Recovery from Massive Faults in Digital Systems", Coordinated Science Laboratory, University of Illinois, Urbana, Quarterly Progress Reports, May 17-November 16, 1970.

719 G. Metze, "Some Problems in Programming Automatic Test Generation Procedures for Digital Networks", *IEEE Intercon 1973*.

720 J. Y. Meuric, P. Pignal and J. Vincent-Carrefour, "Un Programme D'Aide au Control de Reseaux Logiques", *Annales des Telecommunications*, Vol. 26, No. 7, p. 259-77, July 1971.

721 C. S. Meyer, "A Practical Computer Aided System for Logic Simulation and Test Generation", Open Workshop on Fault Detection and Diagnosis in Digital Systems, Lehigh University, 6-8 December 1971.

722 J. F. Meyer and R. J. Sundstrom, "On-Line Diagnosis of Sequential Systems", Workshop on Diagnosis and Reliable Design of Digital Systems, Pasadena, December 5-7, 1973.

723 J. F. Meyer, "Memory Failures in Sequential Machines", University of Illinois, Urbana.

724 J. F. Meyer, "Fault-Tolerant Sequential Machines", *IEEE Transactions on Computer*, Vol. C-20, No. 10, p. 1167-71, October 1971.

725 R. B. Miller, J. D. Foley, and P. R. Smith, "Systematic Troubleshooting and the Half-Split Technique", Technical Report 53-21, Human Resources Research Center, Lackland Air Force Base, July 1953.

726 E. F. Miller and J. H. Pugsley, "Bibliography 22, Sequential Machines", *Computing Reviews*, Vol. 11, No. 5, p. 303-325, May 1970.

727 R. C. Minnick, K. J. Thurber, A. Mukhopadhyay, and K. K. Roy, "Cellular Bulk Transfer Systems, Part A, Fault Location in Cellular Arrays", AD-633 744, October 3, 1968.

728 T. F. Miotke, "Systematic Testing of Boolean Functions", *Computer Design*, March 1972, p. 99-101.

729 G. T. Mitchell, "Problems in Testing Bipolar Memories at the Chip and Card Level", Proceedings of the Symposium-Testing to Integrate Semiconductor Memories into Computer Mainframes, October 4, 1972.

730 A. Miyoshi, "Automatic Generation of Diagnostic Programs for Tosbac 5400-150", 10th Design Automation Workshop, 1973.

731 C. S. Modricker, A. Steen and F. Dintino, "Error Recovery, On-Line Backup, and Maintenance Testing in the Transportable Automatic Digital Switch", Open Workshop on Fault Detection and Diagnosis in Digital Systems, Lehigh University, 6-8 December 1971.

732 D. L. Moon, "Bipolar Logic Test Pattern Generation", *Computer Design*, p. 63-73, March 1971.

733 E. F. Moore, "Gedanken Experiments on Sequential Machines", *Automata Studies*, p. 129-53, Princeton University Press, 1956.

734 V. Moreno, "A Logic Test Generation System Using a Parallel Simulator", Dept. of Computer Science, University of Illinois, Illinois, Illiac IV Document No. 243, February 1971.

735 C. W. Moser, "Checking Wires and Gates in Just One Test Setup," *Electronics*, July 19, 1973, p. 127.

736 M. Moussie and J. Becart, "Recherche D'Une Sequence Pour la Detection des Pannes dans les Circuits Logiques Sequentiels", *L'Onde Electrique*, Vol. 52, No. 3, p. 140-6, March 1972.

737 W. Mow, "Systems Approach to Device Testing", Proceedings of the Symposium-Testing to Integrate Semiconductor Memories into Computer Mainframes, October 4, 1972.

738 B. Mow and B. Mandl, "User Testing of LSI", *Nerem 1973.*

739 F. J. Mowle, "Controllability of Nonlinear Sequential Networks", *Journal of the ACM,* Vol. 17, No. 3, pp. 418-24, July 1970.

740 A. V. Mozgalevskii, "Automatic Fault Detection", Izdat. Mashinostroenie, Leningrad, 1967.

741 D. E. Muller, "Application of Boolean Algebra to Switching Circuit Design and to Error Detection", *IRE Transactions on Electronic Computers,* Vol. EC-13, No. 3, p. 6-11, September 1954.

742 D. E. Muller, "Automatic Fault Diagnosis", University of Illinois, Urbana, 1966.

743 S. Murakami, K. Kinoshita, and H. Ozaki, "Computer Experiments for Finding Fault Detecting Sequences of Sequential Circuits", *Electronics and Communications in Japan,* Vol. 50, No. 10, p. 191-201, October 1967.

744 S. Murakami, K. Kinoshita, and H. Ozaki, "Synthesis of Sequential Machines with Systematic Fault-Detection Sequences", *Electronics and Communications in Japan,* Vol. 51-C, No. 10, p. 112-121, October 1968.

745 S. I. Murakami, K. Kinoshita, and H. Ozaki, "Sequential Machines Capable of Fault Diagnosis", *IEEE Transactions on Computers,* Vol. C-19, No. 11, p. 1079-1085, November 1970.

746 P. E. Murray, "The Scope and Limitations of Testing Harnesses", *Datafair 1973.*

747 G. M. Murzin, "Determination of the Number of Elements Associated with the Group Being Checked When Finding Failures", IZV. Vyssh, Ucheb Zaved Radiotekh, 7, No. 6, p. 697-702, 1964.

748 H. J. Myers and M. Y. Hsiao, "An APL Algorithm for Calculating Boolean Difference", Proceedings of the IEEE Maintainability Conference, Saint-Louis 1968, p. 501-510.

749 F. H. Myers, "Using a Small Digital Computer to Test a Complex Memory System", *NEC 69,* p. 711-14.

750 M. Nagamine, "An Automated Method for Designing Logic Circuit Diagnostic Programs", 8th Design Automation Workshop 1971.

751 J. J. Narraway, "Accessibility in Systems, The Automation of Testing," *IEE Conference Publication No. 91,* p. 1-6, 20-22 September 1972.

752 I. S. Neishtadt, "Method for the Construction of Diagnostic Test for Combinational Logic Circuits", *Automation and Remote Control,* Vol. 9, No. 9, p. 1346-53, September 1967.

753 A. C. Nelson, C. A. Krohn, W. S. Thompson, and J. R. Bats, "Evaluation of Computer Programs for System Performance Effectiveness", Research Triangle Institute, North Carolina, Report RTI-SU-265, 1966.

754 Y. N. Nikolskaya, V. E. Tatur, and B. V. Tikhonova, "Program Localization of Malfunctions in Computer Complexes", AD-762310, May 31, 1973, Edited Translation of Vsesoyuznoe Soveshchanze Po Tekhnicheskoi Diagnostike Trudy, 1972, pp. 22-25.

755 S. Ninomya, T. Hayashi, and M. Nagamine, "Computer-Aided Design for IC Computers", *Fujitsu Scientific and Technical Journal,* March 1969, p. 137-55.

756 K. Nozawa and K. Ritani, "Facom 230-60 Diagnostic Program", 1972 Internation Symposium on Fault-Tolerant Computing, 19-21 June 1972, p. 68-72.

757 J. A. O'Brien, "Computer Fault Location Using Tabular Functions", *IFIP Congress*, Vol. 2, p. 1479-1483, 1968.

758 A. K. Olefir, "The Diagnosis of Digital Systems That Operate in the Presence of Defects", *Vychislitelnye Sistemy*, No. 25, p. 31-43, 1966.

759 A. K. Olefir, "Locating Malfunctioning Elements in a Two-Computer System", *Computer Elements and Systems*, Vol. 1, p. 226-235, 1966.

760 D. C. Opferman and N. T. Tsao-Wu, "On a Class of Rearrangeable Switching Networks, Part I Control Algorithm, Part II Enumeration Studies and Fault Diagnosis," *BSTJ*, Vol. 50, No. 5, May-June 1971, p. 1579-1618.

761 Ya. Ya. Osis, "Minimization of the Number of Test Points", *Automaticheskoe Upravlente*, p. 173-179 (Izdat. Zinatne, Riga, 1967).

762 V. M. Ostianous, "Synthese des Systemes Logiques Facilement Controlables", Colloque International Conception et Maintenance des Automatismes Logiques, Toulouse, September 27-28, 1972.

763 L. A. Ovchinnikov, "An Optimal Method of Detecting the States of a System with Sequential Functional Structure", Trudy Leningradskovo Instituta Aviats, *Priborostroe Niya*, No. 48, p. 108-117, 1966.

764 M. Paige, "Generation of Diagnostic Tests Using Prime Implicants", AD-688 832, May 1969.

765 M. Paige and G. Metze, "Fault Diagnosis in Fet Modules", AD-723 442, May 1971.

766 M. Paige, "Test-Generation Using Prime Implicants", Ph.D. Thesis, Coordinated Science Laboratory, University of Illinois, Urbana, Report R-519, July 1971.

767 M. Paige, "On the Design of Diagnosable Combinational Networks", University of Illinois, Ph.D. Thesis, 1971.

768 M. Paige, "Synthesis of Diagnosable Mosfet", *IEEE Transactions on Computers*, Vol. C-22, No. 5, pp. 513-16, May 1973.

769 M. Paige and G. Metze, "A Design for Diagnosable Combinational Logic", IEEE Computer Society Repository, R-72-242.

770 A. C. Pan, "State Identification and Homing Experiments for Stochastic Sequential Machines", Proceedings of the Third Hawaii Conference, January 1971, pp. 498-500.

771 D. K. Parker, "A Test System for Detecting and Isolating Faults on 4-Phase MOS-LSI PC Boards," *Intercon 1972*, p. 460-61.

772 K. Parker, "Probabilistic Test Generation", 1973 International Symposium on Fault-Tolerant Computing, Palo Alto, June 20-22, 1973.

773 K. Parker, "Probabilistic Test Generation", IEEE Computer Society Repository R73-253.

774 P. P. Parkhomenko, "On Technical Diagnostics", *Znanie 1969*.

775 G. S. Pshkovskii, "Methods of Optimization of Sequential Search for Faults", Izvestiya Akademii Nauk SSSR, *Tekhnicheskaya Kibernetika*, 1971, No. 2, pp. 59-70.

776 J. Passalaqua, C. Massey, and N. Henningham, "Fault Analysis Simulation User's Guide", Information System Division of G. E., October 1969.

777 F. D. Patch and L. M. Zobniw, "Real-Time Diagnosis of Logic Assemblies", 7th Annual Design Automation Workshop, 1970.

778 F. D. Patch and L. M. Zobniw, "Real-Time Diagnosis of Logic Assemblies", Open Workshop on Fault Detection and Diagnosis in Digital Circuits and Systems, Lehigh University, 7-9 December 1970, p. 47-68.

779 W. W. Paterson, "On the Design of Diagnosable Asynchronous Sequential Machines", Ph.D. Thesis, University of Illinois, 1971.

780 W. W. Paterson and G. A. Metze, "A Fault-Tolerant Asynchronous Sequential Machine", 1972 International Symposium on Fault-Tolerant Computing, p. 176-81.

781 N. V. Pautin, "Locating the Malfunctions of Electronic Computers by Programming Means", Diagnostika Neirravnostei Vychislitelnykh Mashin (Izdat. Nauka, Moscow, 1965).

782 B. V. Pavlov, "Cybernetic Methods of Engineering Diagnosis", IZD-VO Mashinostroenie, Moscow, 1968.

783 D. L. Peeler, P. H. Meredith, L. M. Richards, and C. D. Clark, "Automatic Check-Out System for Titan III and Apollo Guidance Computer Programs", IEEE Transactions on Computers, Vol. EC-16, No. 5, p. 580-90, 1967.

784 D. R. Perkins, "Fault Diagnosis to Component Level", Intercon 1972, p. 554-555.

785 M. C. Peterson and R. A. Kirkman, "Quantitative Test Technique for Advanced Automatic Checkout", IRE Transactions on Military Electronics, Vol. MIL-6, p. 240-50, July 1962.

786 A. V. Petrosyan, "Automatic Check of the Functioning of a Digital Computer", Izvestiya A. N. SSSR, Tekhnicheskaya Kibernetika, No. 1, p. 65-72, 1964.

787 K. Pfeuffer, "Algorithmische Verfahren Zur Logischen Prufunc Von Digitalen Schaltnetzen Auf Bestandige Kritische Fehler", AEG-TFK Datenverarbeitung DVP 003 0270, October 1969.

788 K. Pfeuffer, "Pruf-Und Diagnosetests Fur Sequentielle Schaltwerke", AEG-Telefunken, Datenverarbeitung, 1971.

789 K. Pfeuffer, "Error Detection and Diagnosis in Sequential Digital Networks", IEEE Computer Society Repository, R-71-110.

790 T. F. Piatkowski, "Computer Programs Dealing with Finite-State Machines, Part II", AD-658 001, July 1967.

791 P. Pignal,"Fautes Multiples en Reseau Combinatoire", Rapport Interne CNET, November 27, 1968.

792 P. Pignal, G. Roux, and J. Vincent-Carrefour, "Une Methode de Test Automatique Pour les Ensembles Logiques", Onde Electrique, Vol. 48, No. 500, p. 997-1003, et No. 501, p. 1081-88, November and December 1968.

793 P. Pignal, "Detection Automatique des Defauts dans les Circuits Logiques Sequentiels", Colloque Systemes Logiques, Bruxelles, September 1969.

794 J. F. Poage, "Derivation of Optimum Tests to Detect Faults in Combinational Circuits", Digital Systems Laboratory, Princeton University, Technical Report 18, March 1962.

795 J. F. Poage, "The Derivation of Optimum Tests for Logic Circuits", Ph.D. Dissertation, Princeton University, January 1963.

796 J. F. Poage, "Derivation of Optimum Tests to Detect Faults in Combinational Circuits", Proceedings of the Symposium on Mathematical Theory of Automata,

Polytechnic Press, p. 483-528, April 1963.

797 J. F. Poage and E. J. McCluskey, "Derivation of Optimum Test Sequences for Sequential Machines", Digital Systems Laboratory, Princeton University, Technical Report 27, 1964, and 5th Annual Symposium on Switching Theory and Logical Design, p. 121-32, October 1964.

798 Y. S. Pochtar and A. D. Chudakov, "Test System for Static Tests of a Combinative Unit for Comparing Two Quantities Given in Binary Code", *Automation and Remote Control*, Vol. 27, No. 4, p. 678-83, April 1966.

799 E. C. Posner, "Identification of Finite-State Machines", California Institute of Technology, Jet Propulsion Laboratory, Space Programs Summary, 37-32, Vol. IV, p. 256-59, 1965.

800 A. E. Pound, "Designing MOS for Maximum Testability", Wescon Technical Papers, 1971.

801 A. E. Pound, "Testability as a Design Criterion", 1971 IEEE Convention Record, p. 558-59.

802 T. J. Powell, "A Procedure for Ranking Diagnostic Test Inputs", Coordinated Science Laboratory, University of Illinois, Report 354, May 1967.

803 T. J. Powell, "Synthesis Requirements for Fault Detection", Proceedings of the Sixth Allerton Conference on Circuit and Systems Theory, p. 761-72, October 1968.

804 T. J. Powell, "A Procedure for Selecting Diagnostic Test", *IEEE Transactions on Computers*, Vol. C-18, No. 2, p. 168-175, February 1969.

805 T. J. Powell, "A Module Diagnostic Procedure for Combinational Logic", Coordinated Science Laboratory, University of Illinois, Report R.413, and AD-688 743, April 1969.

806 T. J. Powell, "Generating Diagnostic Tests for Modular Combination Logic", IEEE Computer Group Repository, R-70-147.

807 D. K. Pradhan and S. M. Reddy, "Design of Sequential Machines for Reliable Fault Diagnosis," IEEE Computer Society Repository, R73-121.

808 I. V. Prangishvili and V. V. Ignatuschenko, "Methods of Constructing Control—And Diagnostic.Test for Homogeneous Microelectronic Structures", *Engineering Cybernetics*, Vol. 6, No. 11, p. 72-83, 1969.

809 B. A. Prasad and F. G. Gray, "Multiple Fault Detection in Arrays of Combinational Cells", IEEE Computer Society Repository, R73-294.

810 R. J. Preiss, "The Use of Fault-Location Tests in Prototype Bring-Up", Proceedings of the IFIP Congress, Vol. 2, p. 511-512, 1965.

811 R. J. Preiss, "Fault Test Generation", Chapter 7 in *Design Automation of Digital Systems*, Prentice Hall, 1972, M. A. Breuer, Ed.

812 E. M. Prell, "Automatic Trouble Isolation in Duplex Central Controls Employing Matching", *SJCC*, p. 765-70.

813 F. P. Preparata, G. Metze, and R. T. Chien, "On the Connection Assignment Problem of Diagnosable Systems", *IEEE Transactions on Electronic Computers*, Vol. EC-16, No. 6, p. 848-54, December 1967.

814 F. P. Preparata, "Some Results on Sequentially Diagnosable Systems", Proceedings of the First Hawaii International Conference on System Sciences, p. 623-26, January 1968.

815 F. P. Preparata, "An Estimate of the Length of Diagnostic Tests", Third Annual Princeton Conference on Information Sciences and Systems, March 1969.

816 F. P. Preparata, "An Estimate of the Length of Diagnostic Tests", *IEEE Transactions on Reliability*, Vol. R-18, No. 3,pp.131-6, August 1969.

817 F. Prunet, A. Mjara, G. Christol, and G. Urbach, "Contribution Au Test Des Systemes a Relais", Colloque International Sur La Conception Et La Maintenance Des Automates Logiques, Toulouse 1972.

818 F. Prunet and J. M. Dumas, "Verification Des Automates Industriels Par Ordinateur", *Automatisme*, Tome XVII, No. 4, pp. 146-164, April 1973.

819 G. R. Putzolu and J. P. Roth, "An Algorithm and a Program for Generation of Test Patterns for Sequential Circuits", Proceedings of the Third Hawaii International Conference on System Sciences, Honolulu, Western Periodical Company, 1970, p. 64-67.

820 G. Putzolu and J. P. Roth, "A Heuristic Algorithm for the Diagnosis of Asynchronous Circuits", *IEEE Transactions on Computers*, Vol. C-20, No. 6, p. 639-47, June 1971.

821 E. R. Quiet, "An Automated Method for Producing Diagnostic Programs", 1970, 7th Annual Design Automation Workshop.

822 N. N. Raju and P. W. Besslich, "On Fault Equivalence in Combinational Logic Networks", IEEE Computer Society Repository, R79-93.

823 C. V. Ramamoorthy, "The Analytic Design of a Dynamic Look-Ahead and Program Segmenting Scheme for Multiprogrammed Computers", Proceedings ACM 21st National Conference, Thomson Book Co., 1966, pp. 229-39.

824 C. V. Ramamoorthy, "Analysis of Graphs by Connectivity Considerations", *Journal of the ACM*, Vol. 13, No. 2, p. 211-22, April 1966.

825 C. V. Ramamoorthy, "A Structural Theory of Machine Diagnosis," *SJCC 1967*, p. 743-56.

826 C. V. Ramamoorthy, "Computer Fault Diagnosis Using Graph Theory", *Honeywell Computer Journal*, Vol. 1, p. 16-29, Fall 1967.

827 C. V. Ramamoorthy, "The Application of Graph Theory to Computer Diagnosis", Workshop on Hardware Faults and the Programmer, Palm Springs, 28-29 June 1968.

828 C. V. Ramamoorthy and L. C. Chang, "Some Graph Segmentation Procedures for the Parallel Diagnosis of Computer Systems", IEEE International Symposium on Circuit Theory, 14-16 December 1970, p. 35-37.

829 C. V. Ramamoorthy and L. C. Chang, "System Segmentation for the Parallel Diagnosis of Computers", *IEEE Transactions on Computers*, Vol. C-20, No. 3, p. 261-70, March 1971.

830 C. V. Ramamoorthy and W. Mayeda, "Computer Diagnosis Using the Blocking Gate Approach", *IEEE Transactions on Computers*, Vol. C-20, No. 11, pp. 1294-1300, November 1971.

831 C. V. Ramamoorthy and L. C. Chang, "System Modeling and Testing Procedures for Microdiagnostics", 1972 International Symposium on Fault-Tolerant Computing, 19-21 June 1972, p. 110-113.

832 C. V. Ramamoorthy, "Architectural Aspect of Fault-Tolerant Computing", Preprints First Texas Symposium on Computer Systems, June 29-30, 1972.

833 C. V. Ramamoorthy and L. C. Chang, "System Modeling and Testing Procedures for Microdiagnostics", *IEEE Transactions on Computers*, Vol. C-21, No. 11, p. 1169-1183, November 1972.

834 S. Ramos and A. R. Smith, "Fault Detection in Uniform Modular Realizations of Sequential Machines", 1972 International Symposium on Fault-Tolerant Computing, 19-21 June 1972, p. 114-119.

835 H. C. Ratz and B. St. Clair, "A Computer Interface for Fault Diagnosis of Logic Modules", 1971 International Electrical Electronics Conference, p. 74-77.

836 J.C. Rault, "A Graph Theoretical and Probabilistic Approach to the Fault Detection of Digital Circuits", Proceedings of the International Symposium on Fault-Tolerant Computing, 1-3 March 1971, Pasadena, California, p. 26-29.

837 J.C. Rault, "Ecriture des Programmes de Test Automatique", Marche Cri 70027-Rapport Final -March 1971.

838 J.C. Rault, D. Bastin et E. Girard, "Le Test Fonctionnel des Memoires Vives Integrees", Note Technique Thomson-CSF, LCR-DR5, No. 1654, 13 September 1971.

839 J. C. Rault, D. A. Bastin, and E. Girard,"La Detection et la Localisation des Pannes dans les Circuits Logiques - Principles Generaux", *Revue Technique Thomson - CSF*, Vol. 4, No. 1, pp. 49-88, March 1972.

840 J.C. Rault, D. Bastin, and E. Girard, "Une Methode Statistique et Deterministe Pour le Test Fonctionnel des Memoires Vives Integrees", Colloque International Conception et Maintainence des Automatismes Logiques, Toulouse, September 27-28, 1972, also N73-14203.

841 J.C. Rault, E. Girard, and D. Bastin, "La Detection et la Localisation des Defauts dans les Systems Logiques", Masson (A Paraitre 1973).

842 J.C. Rault, E. Girard, and D. Bastin, "La Detection et la Localisation des Defauts dans les Systemes Logiques", Masson (A Paraitre 1974).

843 J. C. Rault, "A Bibliography on Memory Testing", Technical Report No., 1974, Thomson-CSF, 33 Rue de Vouille 75015, Paris, France.

844 C. V. Ravi, "Fault Location in Memory Systems by Program", *SJCC 1969*, p. 393-401.

845 C. V. Ravi, "Fault Location and Graceful Degradation of Memory Systems", Proceedings of the 4th Hawaii International Conference on System Sciences, 12-14 January 1971, p. 736-38.

846 E. M. Raviola, "Logic/Fault Simulator, Error Detection and Correction of Digital Computers—A Bibliography," G. E. Technical Information Exchange, Schenectady, New York, September 16, 1969.

847 V. M. Razin, "Self-Optimized Automatic Fault Location", Izvestiya Tomskovo Politekhnicheskovo Instituta, No. 141, p. 91-2, 1966.

848 V. M. Razin, *et al.*, "Problems in the Automatic Checking of Digital Computer Malfunctions", Izvestiya Tomskovo Politekhnicheskovo Instituta, No. 138, p. 105-107, 1965.

849 S. M. Reddy, "A Design Technique for Fault Locatable Switching Circuit", AD-729 267, June 1971.

850 S. M. Reddy, "Fault Locatable Two Dimensional Cellular Logic Arrays", Sixth Annual Princeton Conference on Information Sciences and Systems, March 23-24, 1972.

851 S. M. Reddy, "Easily Testable Realizations for Logic Functions", Dept. Math., University of Iowa, Iowa City, Themis Report No. 54, May 1972; also AD 745-720.

852 S. M. Reddy, "A Class of Testable Gate Network Realizations for Logic Functions", AD-745 742, May 1972.

853 S. M. Reddy, "Easily Testable Realizations for Logic Functions", 1972 International Symposium on Fault-Tolerant Computing, 19-21 June 1972, p. 126-130.

854 S. M. Reddy, "Easily Testable Realizations for Logic Functions", *IEEE Transactions on Computers*, Vol. C-21, No. 11, p. 1183-88, November 1972.

855 S. M. Reddy, "A Design Procedure for Fault-Locatable Switching Circuits", *IEEE Transactions on Computers*, Vol. C-21, No. 12, p. 1421-1426, December 1972.

856 S. M. Reddy, "Complete Test Sets for Logic Functions", *IEEE Transactions on Computers*, Vol. C-22, No. 11, pp. 1016-1020, November 1973.

857 I. S. Reed, "A Class of Multiple-Error Correcting Codes and the Decoding Scheme", *IRE Transactions on Information Theory*, Vol. IT-4, pp. 38-49, September 1954.

858 R. D. Reese and E. J. McCluskey, "A Gate Equivalent Model for Combinational Logic Network Analysis", 1973 International Symposium on Fault-Tolerant Computing, Palo Alto, California, June 20-22, 1973.

859 K. Reschender and E. Wolf, "Freiprufung Von Koppelstupen", *NTZ* Vol. 24, No. 8, p. 427-32, 1971.

860 G. G. Reynolds, "Test Generation— Simulation System for Digital Circuits," Open Workshop on Fault Detection and Diagnosis in Digital Systems, Lehigh University, 6-8 December 1971.

861 G. G. Reynolds, "A Test Generation—Simulation System for Digital Circuits", Vol. II, Proceedings of the Advanced Digital Technology Conference, Naval Ordinance Laboratory, Silver Springs, Maryland, June 8-10, 1971, p. 53-77.

862 R. Rice, *et al.*, "Illiac IV LSI Memories", *IEEE Journal of Solid State Circuits*, October 1970, pp. 227-228.

863 J. P. Richard, "Detection des Pannes", Rapport LAAS C E 1, March 1971.

864 J. P. Richard, "Contribution a la Detection des Pannes dans les Circuits Logiques", These D'Ingenier—Docteur, Universite Paul Sabatier, Toulouse, September 1971.

865 J. P. Richard, P. Azema, and J. C. Ippolito, "Detection of Multiple Faults in Monotonic Networks—Synthesis of Easily Testable Circuits", IEEE Computer Society Repository, R-72-164.

866 J. P. Richard, "Fonction Booleenne et Test des Pannes Multiples", N72-29177, May 1972, Also LAAS Publication No. 912.

867 J. P. Richard, M. Courvoisier, M. Diaz, and P. Azema, "Detection of Multiple Faults in Combinational Circuits", 1972 International Symposium on Fault-Tolerant Computing, 19-21 June 1972, p. 36-41.

868 J. P. Richard, P. Azema, and J. C. Ippolito, "Detection des Pannes dans un Reseau Monotome-Synthese de Circuits Facilement Testables", Colloque International conception et Maintenance des Automatismes Logiques, Toulouse, September 27-28, 1972.

869 J. P. Richard, "Generation des Tests de Detection des Pannes Multiples Pour Un Circuits Combinatoire", *Automatisme*, Vol. XVII, No. 10, p. 305-317, October 1972.

870 J. P. Richard and P. Azema, "Detection of Multiple Faults in Monotomic Networks and the Synthesis of Easily Testable Circuits", 1973 Fault-Tolerant Computing Symposium, Palo Alto, California, June 1973.

871 D. L. Richards, "Efficient Exercising of Switching Elements in Nets of Identical Gates", *Journal of the ACM*, Vol. 20, No. 1, pp. 88-111, January 1973.

872 D. L. Richards, "Efficient Exercising of Switching Elements in Combinational Nets", *Journal of the ACM*, Vol. 20, No. 2, pp. 320-322, April 1973.

873 R. C. Richmond, "Logical Models of Failure Modes of Digital Integrated Circuits", Workshop on Hardware Faults and the Programmer, Palm Springs, 28-29 June 1968.

874 M. J. Riezenman, "Improved Fault-Finding with Automatic Functional Testers", *Electronics*, August 14, 1972, p. 115.

875 W. B. Riley, *Electronic Computer Memory Technology*, McGraw Hill.

876 E. Rivera, E. R. Garcia, and R. Ranalli, "Computer Generated Fault Isolation Procedures", Proceedings of the Annual Symposium on Reliability, p. 534-41, 1967, Publication IEEE No. 7050.

877 C. Robach, "Synthese et Testabilite D'Une Machine Sequentielle-Adjonction de Sorties Supplementaires", Seminaire D'Algebre Appliquee et Conception, Universite de Grenoble, September 1973.

878 M. A. Robinton, "A Critique of MOS/LSI Testing", *Electronics*, Vol. 44, No. 3, 1 February 1971, p. 62-64.

879 S. M. Ross, "A Problem in Optimal Search and Stop", *Operations Research*, Vol. 17, p. 984-992, 1969.

880 J. P. Roth, "Algebraic Topological Methods in Synthesis", Proceedings of an International Symposium on the Theory of Switching, 2-5 April 1957.

881 I. P. Roth, "Algebraic Topological Methods for the Synthesis of Switching Systems", *Transactions of the American Mathematical Society*, Vol. 88, No. 2, p. 301-26, July 1958.

882 J. P. Roth, "Diagnosis of Automata Failures", Research Memo SR-114, IBM Corp., July 1960.

883 J. P. Roth, "On Computing Diagnostic Tests for Circuits with Feedback", IBM Research Memo, SR-140, IBM Corp., November 29, 1960.

884 J. P. Roth and R. M. Karp, "Minimization Over Boolean Graphs", *IBM Journal of Research and Development*, Vol. 6, No. 2, April 1962.

885 J. P. Roth, "A Pragmatic Theory of Automata, A Symposium On Switching Theory and Automata", IFAC Moscou 1962, and IBM Data Systems Division Report TR.00918, September 21, 1962.

886 J. P. Roth, "Algorithms for the Mechanization of Design", Research Report RC.1294, 12 October 1964.

887 J. P. Roth, "Algorithms for Diagnosis for Automaton Failures", *IBM Journal on Research and Development*, April 1965.

888 J. P. Roth, "Algorithms for Diagnosis of Automata Failures", Research Paper RC-1452, March 30, 1066.

889 J. P. Roth, "Diagnosis of Automata Failures: A Calculus and a Method", *IBM Journal of Research and Development*, Vol. 10, No. 4, p. 278-91, July 1966.

890 J. P. Roth, W. G. Bouricius, and P. R. Schneider, "Programmed Algorithms to Compute Tests to Detect and Distinquish Between Failures in Logic Circuits", *IEEE Transactions on Electronic Computers*, Vol. EC-16, No. 5, p. 567-80, October 1967.

891 J. P. Roth, "On a Method of Design to Facilitate Testing", IBM T. J. Watson Research Center, Report RC 2853, 17 April 1970.

892 J. P. Roth, "An Algorithm to Compute a Test to Distinguish Between Two Failures in a Logic Circuit," IEEE International Computer Group Conference, p. 247-49, 16-18 June 1970.

893 J. P. Roth, "Design of Diagnosable Automata", Proceedings of the ACM Annual Conference, 1972, p. 369.

894 J. P. Roth, "Verify—An Algorithm to Verify a Computer Design, " *IBM Technical Disclosure Bulletin*, Vol. 15, January 1973, pp. 2646-2648.

895 D. M. Rouse and S. A. Szygenda, "Models for Functional Simulation of Large Digital Systems", Digest Record of the Joint Conference on Mathematical and Computer Aided Design, October 1969, p. 363-364.

896 D. M. Rouse, "A Simulation and Diagnosis System Incorporating Various Time Delay Models and Functional Elements", Ph.D. Dissertation, University of Missouri-Rolla, Missouri, 1970.

897 G. C. Rowley, "The Use of Error Correcting Codes in Fault Location", Joint Colloquium on Automatic Aids to Fault Finding in Computers, IERE—IEE, London, February 18, 1965

898 B. K. Roy and H. F. Klock, "An Approach to Compute Diagnostic Test to Detect and Locate Single-Gate Failure in Combinational Digital Circuits", Proceedings IEEE Naecon 1970, p. 226-31.

899 B. K. Roy, "Fault Detection and Diagnosis in Combinatorial Circuits", Ph.D. Thesis, Ohio University, Athens, Ohio, 1971, also AD-754 029, November 1972.

900 B. K. Roy, "Diagnosis and Fault Equivalence in Combinational Circuits", IEEE Computer Society Repository, R73-119.

901 L. Ya. Rozenblyum, "A Method for Removing Failures in Combinational Schemes", *Engineering Cybernetics*, Vol. 9, No. 5, p. 128-31, September-October 1971.

902 J. D. Russell and C. R. Kime, "Structural Factors in the Fault Diagnosis of Combinational Networks", 1971 International Symposium on Fault-Tolerant Computing, et IEEE Transactions on Computers Vol. C-20, No. 11, p. 1276-85, November 1971.

903 J. D. Russell and C. R. Kime, "On the Diagnosability of Digital Systems", 1973 International Symposium on Fault-Tolerant Computing, Palo Alto, California, June 20-22, 1973.

904 J. D. Russell and C. R. Kime, "Structural Design Diagnosability, " Workshop on Diagnosis and Reliable Design of Digital Systems, Pasadena, December 5-7, 1973.

905 R. A. Rutman, "Fault-Detection Test Generation for Sequential Logic by Heuristic Tree Search", IEEE Computer Society Repository, R-72-187.

906 R. A. Rutman, "Fault-Detection Test Generation for Sequential Logic by Heuristic Tree Search", Hughes Aircraft Corp., TP-72-11-4, November 2, 1972.

907 E. Sakrisian, "When is a Component a Component—Ask the Computer", *EDN*, September 2, 1968, p. 32-36.

908 E. Sakrisian, "Large Scale Array Testing Techniques", *NEREM*, 1969, p. 220-21.

909 A. B. Salisbury and P. H. Enslow, "Diagnostic Programming for Digital Computers—A Bibliography," West Point Military Academy, AD-813 831, April 1967.

910 A. B. Salisbury and P. H. Enslow, "Diagnostic Programming for Digital Computers", *Computing Reviews*, Vol. 9, No. 9, September 1968, p. 573-575.

911 K. K. Saluja and S. M. Reddy, "On Minimally Testable Logic Networks", IEEE Computer Society Repository, R73-31.

912 K. K. Saluja and S. M. Reddy, "Fault Detecting Test Sets for Reed-Muller Canonic Networks", IEEE Computer Society Repository, R73-109.

913 K. K. Saluja and S. M. Reddy, "Easily Testable Two Dimensional Cellular Logic Arrays", IEEE Computer Society Repository, R73-196.

914 K. K. Saluja and S. M. Reddy, "Multiple Faults in Reed-Muller Canonic Networks", Proceedings of the 1972 Switching and Automata Theory, pp. 185-191.

915 S. Salvo and C. Levy, "Detection des Defauts Logiques dans la Logique Synchrone", SESA, RS166/No. 9203/69.

916 B. M. Saper, "A Fresh Look at Comparative Troubleshooting Techniques for the Maintenance of Digital Equipment", Automatic Support Systems Symposium for Advanced Maintainability, November 13-15, 1972, p. 9-16.

917 A. A. Sarris and M. Mor, "Sequential Circuit Modeling for Fault Test Generation", IEEE Computer Society Repository R73-286.

918 G. Saucier and M. A. Verdillon, "Structures Facilement Testables", Seminaire D'Algebre Appliquee et Conception, Universite de Grenoble, February 28, 1972.

919 G. Saucier, "Test de Systemes Sequentiels", Seminaire Imag., March 13, 1972.

920 G. Saucier, "Recherche D'Une Sequence de Test D'Une Machine Sequentielle", *Rairo*, Vol. 6, No. 1, p. 99-105, 1972.

921 G. Saucier, "Structures Facilement Testables-Reseau Ni a Parite", Colloque International Conception et Maintenance des Automatismes Logiques, Toulouse, September 27-28, 1972, also N73-12258.

922 G. Saucier, "Computer Test Programs and Vector Tests for Large Components", Workshop on Diagnosis and Reliable Design of Digital Systems, Pasadena, December 5-7, 1973.

923 D. L. Sauder, "Cut Wiring System Tests Exponentially by Using Binary Group Checks to Indicate and Locate Faults, It is Easier with a Computer", *Electronic Design*, Vol. 16, No. 3, p. 94-99, February 1968.

924 Yu. G. Savchenko, "Some Problems of Fault Detection in Redundant Automata", SB. Avtomatizatsiya Proektirovanie, Tekhnologiya i Programmirovannoe Upravlenie, pp. 149-156, Izdat Tekhnika, Kiev, 1966.

925 D. H. Sawin, G. K. Maki, and S. R. Groenig, "Design of Asynchronous Sequential Machines for Fault Detection", 1972 International Symposium on Fault-Tolerant Computing, p. 170-75.

926 D. H. Sawin and G. K. Maki, "Asynchronous Sequential Machines Designed for Fault Detection", IEEE Computer Society Repository, R73-129.

927 B. Schallop, "Error Trees and Calculation Rules for the Failure Performance of Logic Circuits", *Int. Elektron. Rundsch.*, Vol. 25, No. 1, pp. 7-10, January 1971.

928 B. H. Scheff and S. P. Young, "Fault Simulation in Design Automation of Digital

Systems", Chapter 3, p. 143-150, M. A. Breuer Ed., Prentice Hall 1972.

929 F. Schenstrom and R. Williams, "Computer Aided Design MOS", *Electronic Engineer*, Vol. 29, No. 3, p. 70, March 1970.

930 D. R. Schertz and G. A. Metze, "On the Indistinguishability of Faults in Digital Systems" Proceedings of the 6th Annual Allerton Conference on Circuits and System Theory, 1968, p. 752-60.

931 D. R. Schertz, "On the Representation of Digital Faults", AD-688 836, May 1969.

932 D. R. Schertz and G. A. Metze, "The Use of Connection Graphs for the Detection of Digital Faults", Proceedings of the 7th Allerton Conference on Circuit and System Theory, 1969, pp. 262-71.

933 D. R. Schertz and G. A. Metze, "On the Design of Multiple Fault Diagnosable Networks", *IEEE Transactions on Computers*, Vol. C-20, No. 11, p. 1361-64, November 1971.

934 D. R. Schertz and G. A. Metze, "Comment on Derivation of Minimal Complete Sets of Test-Input Sequences Using Boolean Differences", IEEE Computer Society Repository R-71-176.

935 D. R. Schertz and G. A. Metze, New Representation for Faults in Combinational Digital Circuits", *IEEE Transactions on Computers*, Vol. C-21, No. 8, p. 858-866, August 1972.

936 D. R. Schertz, "Applications of a Fault Class Representation to Diagnosis", Third International Symposium on Fault-Tolerant Computing, Palo Alto, June 20-22, 1973.

937 D. R. Schertz, "On the Diagnosis of Multiple Output Circuits", Eleventh Annual Allerton Conference on Circuit and System Theory, October 3-5, 1973.

938 P. Schlemmer, "Tester Programmed by Responses of Test Device to Prior Interrogation", *IBM Technical Disclosure Bulletin*, Vol. 14, No. 5, p. 1419-21, October 1971.

939 G. Schnable and R. Keen, "Failure Mechanisms in Large Scale Integrated Circuits", *IEEE Transactions on Electronic Devices*, Vol. Ed, No., p. 322-332, April 1969.

940 P. R. Schneider, "On the Necessity to Examine D-Chains in Diagnostic Test Generation—An Example", *IBM Journal of Research and Development*, Vol. 11, No. 1, p. 114, January 1967.

941 S. Schneider and D. H. Wagner, "Error Detection in Redundant Systems", Proceedings on the WJCC, p. 115-21, 26-28 February 1957.

942 R. Schroder, "Fault Trees for Reliability Analysis".

943 D. Schulte, "Maintenance of TSPS Common Control Equipment", Open Workshop on Fault Detection and Diagnosis in Digital Systems, 6-8 December 1971, Lehigh University.

944 P. J. Scola, "On-Line Computer System Testing", Open Workshop on Fault Detection and Diagnosis in Digital Systems, Lehigh University, 6-8 December 1971.

945 P. J. Scola, "Logic Design Guide Lines for Maintainability", *Intercon 1972*, p. 46-47.

946 P. J. Scola, "An Annotated Bibliography of Test and Diagnostic", *Honeywell Computer Journal*, Vol. 6, No. 2, p. 97-161, 1972.

947 C. L. Seitz, "An Approach to Designing Checking Experiments Based on a Dynamic

Model", *Theory of Machines and Computation*, Z. Kohavi and A. Paz, p. 341-349, Academic Press 1971.

948 F. F. Sellers, L. W. Bearnson, and M. Hsia, "Analyzing Errors with the Boolean Difference", Digest of the First Annual IEEE Computer Conference, 6-8 September 1967.

949 F. F. Sellers, M-Y, Hsiao, and L. W. Bearnson, "Error Detecting Logic for Digital Computers", Chapter 14, *Memory and Storage Error Detection*, McGraw Hill, 1968.

950 F. F. Sellers, L. W. Bearnson and M. Y. Hsiao, "Analyzing Errors with the Boolean Difference", *IEEE Transactions on Electronic Computers*, Vol. EC-17, No., p. 676-83, July 1968, *Errata in IEEE Transactions on Electronic Computers*, Vol. C-18, No. 2, p. 381, April 1969.

951 F. F. Sellers, M. Y. Hsiao, and L. W. Bearnson, *Error Detecting Logic for Digital Computers*, McGraw Hill, 1969.

952 S. N. Semanderes, "Elraft, A Computer Program for the Efficient Logic Reduction Analysis of Fault Trees", *Transactions of the American Nuclear Society*, Vol. 13, No.2, p. 781-2, 1970, *Et IEEE Transactions on Nuclear Science*, Vol. NS-18, No. 1, p. 481-86, February 1971.

953 S. Seshu, "The Diagnosis of Asynchronous Switching Systems", Conference on the Diagnosis of Failures in Switching Circuits, Michigan State University, East Lansing, May 1961.

954 S. Seshu and D. N. Freeman, "The Diagnosis of Asynchronous Sequential Switching Systems", *IRE Transactions on Electronic Computers*, Vol. EC-11, No. 4, p. 459-65, August 1962.

955 S. Seshu, "The Future of Diagnosis", Proceedings of the Seminar on Automatic Checkout Techniques, Battelle Memorial Institute, 5-7 September 1962, p. 127-31.

956 S. Seshu, "On an Improved Diagnosis Program", *IEEE Transactions on Electronic Computers*, Vol. EC-14, No., p. 76-79, February 1965 and AD-601 156, May 1964.

957 S. Seshu, "The Logic Organizer and Diagnosis Programs", AD-605927, July 1964.

958 S. C. Seth, "Fault Diagnosis of Combinational Cellular Arrays", Proceedings of the 7th Allerton Conference on Circuit and System Theory, p. 272-83, October 1969.

959 S. C. Seth , "Distance Measures on Detection Test Sets and Their Application", 1973 International Symposium on Fault-Tolerant Computing, Palo Alto, California, June 20-22, 1973.

960 S. C. Seth, "Distance Functions on Detection Test Sets in Combinational Networks", IEEE Computer Society Repository, R73-246.

961 H. G. Shab, B. D. Carroll, and D. M. Junes, "Multiple Faults in Combinational Logic," AD-758 165, March 1973.

962 G. Shapiro, G. J. Rogers, O. B. Lang, and P. M. Fulcomer, "Project Fist-Fault Isolation by Semiautomatic Techniques", *IEEE Spectrum*, Vol. 1, No., p. 98-111, August 1964, Et p. 130-144, September 1964.

963 M. S. Shcherbinin and B. F. Ivanov, "The Probability Forecasting of Malfunction Points in Computers", AD-762 263, June 1, 1973, Edited Translation of Vsesoyuznoe Soveshchanze po Tekhnicheskoi Diagnostike Trudy, 1972, pp. 64-66.

964 D. A. Sheppard and A. G. Vrasenic, "Fault Detection of Binary Sequential Machines Using N-Valued Test Machines", IEEE Computer Society Repository, R73-126.

965 R. A. Short and J. Goldberg, "A Summary of Soviet Activities in the Design of Fault-Tolerant Digital Machines", *Computer*, January/February 1971, pp. 28-33.

966 R. A. Short and J. Goldberg, "Soviet Progress in the Design of Fault-Tolerant Digital Machines", *IEEE Transactions on Computers*, Vol. C-20, No. 11, p. 1337-52, November 1971.

967 H. Shoyo, K. Tadao, and O. Hiroshi, "Some Consideration on the Testing of Logic Circuits", Electronic Computer Research Committee, January 1963.

968 R. W. Shrader, "A Logic Design Procedure to Facilitate Fault Detection", Report R-513 (AD-726 382), University of Illinois, Urbana, Illinois, June 1971.

969 V. M. Shukla and R. G. Burke, "Computer-Aided Testing and Diagnosis of Integrated Circuits—I", Collins Radio Company, April 25, 1968.

970 V. M. Shukla, "Computer-Aided Testing and Diagnosis of Integrated Circuits—II", Collins Radio Company, October 1, 1968.

971 V. M. Shukla, "Computerized Logic Simulation, Test Verification and Fault Isolation of Digital Circuits", Northrop Corporation, December 19, 1970.

972 V. M. Shukla, "Functional Testing of Complex Logic Circuits", Symposium on Hybrid Microelectronics, 1971, pp. 9-5-1 to 9-5-9.

973 D. P. Siewiorek, "An Improved Algorithm for Selecting a Set of Diagnostic Tests", Digital Systems Laboratory, Stanford Electronic Laboratories, Stanford University, Technical Report No. 23, December 1971.

974 I. M. Sindeyev, "Synthesizing Logical Schemes for Failures Detection and Control of Complex Systems", *Engineering Cybernetics*, No. 2, p. 16-23, March-April 1963.

975 S. Singh and V. P. Pingh, "A Probabilistic Approach for Testing Circuits", IEEE Computer Society Repository, R-72-183.

976 A. N. Sklyarevich, "Function for Testing the Working Order of a Logic Element in a Combinational Automat", *Vychislist Tekh*, No. 3, p. 55-61, 1970.

977 A. N. Sklyarevich, "Possibilite de Verification D'Un Test D'Automate Combinatoire", Automat Vychislit, Tekh. Latv. SSR, No. 2, p. 20-25, 1971.

978 A. N. Sklyarevich, "Caracheristiques D'Une Suite Complete de Tests D'Un Automate Combinatoire", Automat. Vychislit. Tekh., Latv. SSR, 1971, 5, No. 4, p. 17-25.

979 A. N. Sklyarevich, "Tests for a Complete Logic Check of a Noniterative Combinatorial Automaton", *Automation and Remote Control*, No. 1, 1972.

980 E. D. Smally, "Computer-Controlled Software-Diagnosis of an Airborne Computer", Digest of Papers of the Sixth Annual IEEE Computer Society International Conference, San Francisco, 12-14 September 1972, pp. 231-3.

981 E. D. Smally, "Computer-Controlled Diagnostic of Another Computer", Automatic Support Systems Symposium for Advanced Maintainability, November 13-15, 1972.

982 A. S. Smirnov, "Informational Reliability of Electronic Combinational Circuits", *Automation and Remote Control*, Vol. 33, No. 8, pp. 1382-90, 1972.

983 E. W. Smith and B. D. Carroll, "A Computer Aid for Research and Teaching in Fault Diagnosis of Logic Networks", Proc. IEEE Region 3, 1972.

984 E. W. Smith and B. D. Carroll, "A Computer Aid for Research and Teaching in Fault Diagnosis of Logic", Proc. 10th Region 3 Convention, 1973.

985 G. R. Smith and S. S. Yau, "A Programmed Fault-Detection Algorithm for

Combinational Switching Networks", Proceedings NEC 1969, p. 668-73.

986 G. W. Smith and S. G. Chappell, "Machine Aids for Logic Design Verification, Fault Detection, and Fault Isolation", 1971 IEEE International Computer Conference, p. 63-64.

987 G. W. Smith and S. G. Chappell, "Machine Aids for Logic Design Verification, Fault Detection, and Fault Isolation", Intercon 1972, p. 56-57.

988 G. W. Smith, "Fault Detection Test Derivation Using Boolean Difference Techniques", Proceedings of the ACM Annual Conference, 1972, p. 369-78.

989 W. R. Smith, "Digital Diagnostic Techniques, Survey, and Recommendations", AD-696 932, October 1969.

990 T. J. Snethen and D. C. Jessep, "The Circuits Failure Modelling Challenge for LSI", Proceedings of the 1973 IEEE International Symposium on Circuits Theory, Toronto, April 9-11, 1973.

991 E. S. Sogomonyan, "Monitoring Operability and Finding Failures in Functionally Connected Systems", *Automation and Remote Control*, Vol. 25, No. 6, p. 874-82, June 1964.

992 E. S. Sogomonyan, "A Device for Searching for Defective Units and Elements of Functionally Connected Systems, Bulletin Izobretenit i Tovarnykh Znakov No. 24, p. 105-106, 1965, (USSR Scientific Abstracts, CCAT, No. 18, p. 81).

993 E. S. Sogomonyan, "Problems of the Analysis of Combinational Multiterminal Networks for Purposes of Testing the Operative State and Locating Faults", *Abstracts and Structural Theory of Switching Devices*, M. Gavriloved., p. 225-241, Izdat. Nauka, Moscow, 1966.

994 E. S. Sogomonyan, " Failure Diagnosis in Discrete Block Objects", *Automation and Remote Control*, October 1969.

995 E. S. Sogomonyan, "The Construction of Discrete Objects with Diagnosis During the Operation", *Automation and Remote Control*, November 1970.

996 N. A. Solovev, "Tests for a Certain Class of Planar and Spatial Tables", International Congress of Mathematicians, Abstracts of Short Scientific Communications, No. 13, p. 30, (Izdatelstvo Nauka, Moscow 1966).

997 R. G. South, "A System for Simulating Faults in Large Logic Circuits", Open Workshop on Fault Detection and Diagnosis in Digital Systems, Lehigh University, 6-8 December 1971.

998 D. J. Spencer, "Identification of Discrete State Systems", Ph.D. Thesis, Dept. of Computer Science, UCLA, 1970.

999 M. A. Spivak, "Generalized Diagnosis and Setup Problems for Finite Automata", *Engineering Cybernetics*, Vol. 8, No. 3, p. 72-76, May-June 1969.

1000 C. V. Srinivasan, "Codes for Error Correction in High Speed Memory Systems, Part I Correction of Cell Defects in Integrated Memories", *IEEE Transactions on Computers*, Vol. C-20, No. 8, p. 882-888, August 1971.

1001 W. R. Stanley, "An Optimal System Design for Printed Circuit Board Test Reporting", *Computer Design*, January 1970, p. 73.

1002 E. St. Clair, "The Diagnosis Proceedings of the Seminar on Automatic Checkout Techniques", Battelle Memorial Institute, Columbus, Ohio, 5-7 September 1962.

1003 S. K. Stearns, "Store Diagnostic Development Using Digital Simulation", IEEE Computer Group Repository R-70-102, 25 March 1970.

1004 D. Stein, "Fault Isolation for Digital Testers", IEEE 1974 Intercon.

1005 P. D. Stigall and T. D. Steury, "Connectivity Matrix Approach to Fault Equivalence in Combinational Logic Networks", IEEE Computer Society Repository, R-72-197.

1006 P. D. Stigall and T. D. Steury, "Connectivity Matrix Approach to Fault Equivalence in Combinational Logic Networks", Proceedings of the IEEE Region III Convention, 1973.

1007 G. N. Stockwell, "Computer Logic Testing by Simulation", *IRE Transactions on Military Electronics*, Vol. MIL-6, No. 3, p. 275-82, July 1962.

1008 L. M. Stolurow, B. Bergum, T. Hodgson, and J. Silva, "The Efficient Course of Action in Troubleshooting as a Joint Function of Probability and Cost", *Educational and Psychological Measurement*, Vol. 15, No. 4, p. 462-77, Winter 1955.

1009 A. M. Stoughton, "Computer-Controlled Memory Testing", *Modern Data Systems*, August 1968.

1010 A. M. Stoughton and R. J. Merckert, "Computer-Controlled Testing for Improving Ferrite Core Memory Design", *IEEE Transactions on Magnetics*, Vol. MAG-5, No., p. 651-656, September 1969.

1011 J. E. Stuehler, "Hardware-Software Tradeoffs when Applying Computers to Testing", *Wescon*.

1012 S. Y. H. Su and Y.C. Cho, "A New Approach to the Fault Location of Combinational Circuits," *IEEE Transactions on Computers*, Vol. C-20, No. 1, p. 21-30, January 1972.

1013 S. Y. H. Su, S. J. Chang, and M. Brouer, "Location of Multiple Stuck-Type Faults in Combinational Networks", IEEE Computer Society Repository, R-72-196.

1014 C. H. Sung, "Fault Detection and Location in Sequential Cellular Arrays", University of Texas, Austin, Ph.D. Thesis, 1971.

1015 C. H. Sung and C. L. Coates, "Fault Detection and Location in Sequential Cellular Arrays", AD-736 765, July 5, 1971.

1016 C. H. Sung and C. L. Coates, "Tessallation Aspect of Fault Detection and Location in Combinational Cellular Arrays", IEEE Computer Society Repository, R73-66.

1017 A. K. Susskind, "Additional Applications of the Boolean Difference to Fault Detection and Diagnosis", 1972, International Symposium on Fault-Tolerant Computing, 19-21 June 1972, p. 58-61.

1018 A. K. Susskind, "Diagnosis for Logic Networks", *IEEE Spectrum*, October 1973, pp. 40-47.

1019 S. A. Szygenda and G. C. Goldbogen, "Implementation and Extension of Multidimensional Path Sensitizing in a Simulation and Diagnosis System", Proceedings of the 7th Allerton Conference on Circuit and System Theory, 1969, p. 284-92.

1020 S. A. Szygenda and D. Rouse, "Functional Models for Simulation of Large Digital Systems", Proceedings of the Joint Conference on Mathematical and Computer Aids to Design, October 1969.

1021 S. A. Szygenda, "A Software System for Diagnosis Test Generation and Simulation of Large Digital Systems", *NEC 1969*, p. 657-62.

1022 S. A. Szygenda, "Tegas-A Diagnostic Test Generation and Simulation System for Digital Computers", Proceedings of the 3rd Hawaii International Conference on System Sciences, January 1970.

1023 S. A. Szygenda and D. M. Rouse, "Functional Simulation—A Basis for a Systems Approach to Digital Simulation", IEEE Computer Society Repository R-70-93, 24 February, 1970.

1024 S. A. Szygenda and M. J. Flynn, "Failure Analysis of Memory Organization for Utilization in a Self-Repair Memory System", *IEEE Transactions on Reliability*, Vol. R-20, No. 2, p. 64-70, May 1970.

1025 S. A. Szygenda, D.M. Rouse, and E. W. Thompson, "A Model and Implementation of a Universal Time-Delay Simulator for Large Digital Nets", *SJCC 1970*, p. 207-16.

1026 S. A. Szygenda, "Problems Associated with the Implementation and Utilization of Digital Simulators and Diagnostic Test Generation Systems", International Symposium on Fault-Tolerant Computing, March 1971, pp. 51-53.

1027 S. A. Szygenda and M. J. Flynn, "Coding Techniques for Failure Recovery in a Distributive Modular Memory Organization", *SJCC 1971*, p. 459-66.

1028 S. A. Szygenda, C. W. Hemming, and J. M. Hemphill, "Time Flow Mechanisms for Use in Digital Logic Simulation", Proceedings of the 5th Annual Conference on Application of Simulation, December 1971.

1029 S. A. Szygenda, "A Simulator for Digital Design and Diagnosis," Open Workshop on Fault Detection and Diagnosis in Digital Systems, Lehigh University, 6-8 December 1971.

1030 S. A. Szygenda, J. Hemphill, C. Hemming, J. Fike, A. Lekkos, and E. Thompson, "Implementation of Synthesized Techniques for a Comprehensive Digital Design Verification and Diagnosis System", Proceedings of the 5th Hawaii International Conference on System Science, Honolulu, Hawaii, 11-13 January 1972, p. 293-5.

1031 S. A. Szygenda, "Tegas 2—Anatomy of a General Purpose Test Generation and Simulation for Digital Logic", Proceedings of the 9th ACM-IEEE Design Automation Workshop, June 1972, p. 116-127.

1032 S. A. Szygenda, C. Hemming, and D. M. Rouse, "Functional Simulation—A Basis for a Systems Approach to Digital Simulation and Fault Diagnosis", 1972 Summer Simulation Conference, June 1972, p. 270-281.

1033 S. A. Szygenda and E. W. Thompson, "Fault Insertion Techniques and Models for Digital Logic Simulation", *FJCC 1972*, pp. 875-884.

1034 K. Tadoa, H. Shoyo, and O. Hiroshi, "Some Considerations on the Determination of Fault Location in Combined Logic Circuits", *Research on Electronic Computers*, 2nd Issue 1962.

1035 M. Takesue, "Relations Between Diagnostic Resolution and Structure of Logical Circuits", *Elec. Comm. Lab. Tech. J.*, Vol. 21, No. 2, p. 291-301, 1972.

1036 E. Tammaru and J. B. Angell, "Redundancy for LSI Yields Enhancement", *IEEE Journal on Solid State Circuits*, Vol. SC-2, No. 4, p. 172-82, December 1967.

1037 E. Tammaru, "Efficient Testing of Combinational Logic Cells in Large Scale Arrays", Ph.D. Dissertation, Stanford Univeristy, Stanford Electronic Laboratory, Technical Report No. 4601-1, SU-SEL-68-040, May 1968.

1038 E. Tammaru, "Testing of Combinational Logic Cells in Large Scale Arrays", IEEE Computer Group Repository R-69-115, 26 May 1969.

1039 C. Tanaka, K. Tabuchi, and H. Kanada, "A Procedure for Selecting Diagnostic Tests in Combinational Circuits", IEEE Computer Society Repository, R-70-171.

1040 J. C. Tarbouriech, "Methodes de Determination de Sequences de Test Pour les Circuits Logiques Combinatoires et Sequentiels", Rapport Interne Sescosem.

1041 R. Tellez-Giron, "Une Machine a Tester les Systemes Sequentiels par des Sequences Pseudo-Aleatoires", Colloque International Conception et Maintenance des Automatismes Logique, Toulouse, September 27-28, 1972, also N73-12196.

1042 A. Thayse, "Transient Analysis of Logical Networks Applied to Hazard Detection", Philips Research Report, Vol. 25, No. 5, p. 261-336, October 1970.

1043 A. Thayse, "Boolean Differential Calculus", Philips Research Report, Vol. 26, p. 229-46, June 1971.

1044 A. Thayse, "A Variational Diagnosis Method for Stuck-Faults in Combinatorial Networks", Philips Research Report, Vol. 27, No. 1, p. 82-98, February 1972.

1045 A. Thayse, "Testing of Asynchronous Sequential Switching Circuits", Philips Research Report, Vol. 27, No. 1, p. 99-106, February 1972.

1046 A. Thayse, "Multiple-Fault Detection in Large Logical Networks", Philips Research Report, Vol. 27, pp. 583-602, 1972.

1047 A. Thayse and M. Davio, "Boolean Differential Calculus and its Application to Switching Theory", IEEE Transactions on Computers, Vol. C-22, No. 4, pp. 409-420, April 1973.

1048 A. Thayse and J. P. Deschamps, "Derivatives of Discrete Functions and Their Application to Switching Theory", IEEE Computer Society Repository, R73-204.

1049 J. Thomas, "Automated Diagnostic Test Programs for Digital Networks", Computer Design, Vol. 10, No. 8, p. 63-67, August 1971.

1050 I. Thomas, Establishing Test Generation Requirements for Digital Networks", Wescon 1973.

1051 E. W. Thompson, "The Analysis and Synthesis of Methods for the Modeling of Different Fault Classes in a Digital Logic Simulation System", Southern Methodist University, Dallas, Texas, May 1972.

1052 E. W. Thompson and S. A. Szygenda, "Specification and Generation of Different Fault Classes for Digital Simulation", 1973 Southwestern IEEE Conference, pp. 388-95.

1053 K. J. Thurber, "Fault Location in Cellular Arrays", FJCC 1969, p. 81-88, and AD-683 744, October 3, 1968.

1054 L. S. Timonen, "The Construction of Optimal Programs for Diagnosing the State of Complex Engineering Systems", Engineering Cybernetics, No. 4, p. 94-100, 1966.

1055 L. S. Timonen, "The Construction of Optimal Programs for Checking Operability", Autometriya, No. 1, 1966.

1056 L. S. Timonen, "The Application of Graph Theory to the Analysis of Systems of Engineering Diagnostics", Proceedings of the Seventh All-Union Conference on Automatic Testing and Methods of Electrical Measurements (Izdat. Sibirskoe Otdelnie, A. N. SSSR, Novosibirsk, 1967).

1057 K. To, "Fault Folding, A Unified Approach to Fault Analysis in Combinational Logic Networks", 1971 IEEE Computer Society Conference, p. 81-82.

1058 K. To, "Fault Folding for Irredundant and Redundant Combinational Circuits", IEEE Transactions on Computers, Vol. C-22, No. 11, pp. 1008-1015, November 1973.

1059 Y. Tohma, "Design Technique of Fail-Safe Sequential Circuits Using Flip-Flops for Internal Memory", IEEE Computer Repository, R73-287.

1060 Y. L. Tomfeld, "The Diagnostics of Malfunctions Which Cause Competitions", AD-762234, June 8, 1973—Edited Translation of Vsesoyuznoe Soveshchanie po Tekhnicheskoi Diagnostike Trudy, 1972, pp. 228-232.

1061 J. R. Townes, S.J. Dwyer, and G. B. Zobrist, "Fault Identification by Computer", Electrotechnology, April 1968, p. 112, 115.

1062 J. J. Travalent, "Fault Catalog Preparation Through Simulation Methods", Computer Designer's Conference and Exhibition, Anaheim, California, 19-21 January 1971.

1063 J. J. Travalent, "Fault Isolation and Cataloguing", Computer Designer's Conference, Anaheim, California, January 1971, pp. 51-55.

1064 S. H. Tsiang and W. Ulrich, "Automatic Trouble Diagnosis of Complex Logic Circuits", BSTJ, Vol. 41, p. 1170-1200, July 1962.

1065 S. H. Tsiang and W. Ulrich, "Automatic Trouble Diagnosis of Complex Logic Circuits", Transactions on Communications and Electronics, AIEE, January 1963.

1066 T. T. Tylaska, "Fault Detection in Sequential Machines", AD-750 266, September 14, 1972.

1067 T. T. Tylaska and J. D. Bargainer, "An Improved Bound for Checking Experiments That Use Simple Input-Output and Characterizing Sequences", IEEE Computer Society Repository, R73-60.

1068 E. G. Ulrich, "The Evaluation of Digital Diagnostic Programs Through Digital Simulation", Computer Technology, IEEE Conference Publication No. 32, p. 9-19, 1967.

1069 E. G. Ulrich, T. Baker and L. R. Williams, "Fault-Test Analysis Techniques Based on the Use of Logic Simulation", 9th Annual Design Automation Workshop, Dallas, 26-28 June 1972, p. 111-15.

1070 E. G. Ulrich and T. Baker, "The Concurrent Simulation of Nearly Identical Digital Networks", 10th Design Automation Workshop 1973.

1071 R. W. Ulrickson, "Testing of Complex Integrated Logic Circuits", Nerem Record, 1967, p. 136-137.

1072 Y. Umetani, S. Miyamoto, and H. Horikoshi, "Some Results on the Application of FLT Generation to a Large Scale Computer", 1973 International Symposium on Fault-Tolerant Computing, Palo Alto, California, June 20-22, 1973.

1073 S. H. Unger, "Fault Checking of Combinational Circuits with Input Reconvergent Fan-Out", Colloque International Conception et Maintenance des Automatismes Logiques, Toulouse, 27-28 September 1972, also N73-12257.

1074 D. D. Urban, "Digital Computer Testing, Dans Electronic Testing", L. Farkas, Chapitre 14, p. 275-93, McGraw Hill, New York, 1966.

1075 Uvarova, "Automatic Programmed Control of Computers in Computer Fault Diagnosis", IZD. Nauka 1965.

1076 V. V. Vaksov, "On Test for Irreversible Switching Circuits", Automation and Remote Control, Vol. 26, No. 3, p. 515-18, March 1965.

1077 F. R. Valette and G. Gelis, "Diagnostic Automatique des Pannes dans les Systems Informatiques", Centre D'Informatique de Toulouse, October 1968.

1078 F. R. Valette, "Detection et Diagnostic des Pannes dans les Systems de Logique Combinatoire", Colloque Microelectronique Toulouse, 1969.

1079 R. P. Vancura and C. R. Kime, "On Numerical Bounds in Diagnosable Systems", 1972 International Symposium on Fault-Tolerant Computing, 19-21 June 1972, p. 148-153.

1080 A. Van de Grijp, T. Holtwijk, and R. M. G. Wijnhoven, "Satan—A Versatile Array Tester", IEEE Transactions on Magnetics, Vol. MAG-5, pp. 656-661, September 1969.

1081 K. L. Vange, "Obtaining the Initial Data for the Diagnosis of Finite Automata", Avtomatika i Vychislitelnaya Tekhnika, Vol. 2, No. 1, p. 31-36, 1968.

1082 K. L. Vange, "Optimal Checking in Combinational Automata", Automation and Remote Control Engineering, Vol. 2, No. 5, p. 48-52, 1968.

1083 K. L. Vange, "Construction of a Diagnostic Test Sequence for a Combined Automaton, in Problems of Synthesis of Finite Automata (V.I. Levin, ed.), Riga, USSR, Izdatel' Stvo Izinatne', 1972.

1084 M. M. Vartanian, "An Algorithm for Fault Isolation of Multi-State Electronic Networks", Ph.D. Dissertation, University of Pennsylvania, 1969.

1085 A. G. Varzhapetyan and Yu. E. Sazhin, "On Acceleration of the Process of Statistical Simulation of the Failures of Elements of a System", Automatika i Vychislitelnaya Tekhnika, 1971, No. 2, pp. 95-96.

1086 A. V. Vasilenskii, "Signal Transmission Diagnostics and Its Use in Autonomous Relay Systems", Nauchno-Tekhnicheskiisbornik Institut Tekhnicheskoi Kibernetiki, A. N. BSSR, p. 132-38, 1968.

1087 M. J. Vaughn, "Technical and Economic Aspects of Core Plane Testing", Computer Design, Vol. 12, No. 1, p. 79-82.

1088 V. A. Vedeshenkov, "Construction of Test Tables for Detecting Logical Errors of Electronic Combinational Devices", Automation and Remote Control, Vol. 23, No. 3, p. 491-500, March 1968.

1089 A. Verdillon, "Structures sur L'Ensemble des Pannes Multiples D'Un Reseau Combinatoire", Colloque sur la Conception et la Maintenance des Automatismes Logiques, Toulouse, September 1972, also N73-13247.

1090 A. Verdillon, "Adjonction de Points de Test—Recouvrement de Reseaux par des Arbres", Colloque International sur la Conception et la Maintenance des Automatismes Logiques, Toulouse, September 1972.

1091 A. Verdillon, "Failures in Acyclic Networks", 1973 Fault-Tolerant Computing Symposium, Palo Alto, June 19-22, 1973.

1092 A. Verdillon, "Pannes dans les Reseaux Acycliques", These de Troisieme Cycle, Universite de Grenoble, December 1972.

1093 G. F. Verzakov and L. S. Timonen, "Minimization of the Symptoms for Fault Detection", Proceedings of the All-Union Conference on Automatic Testing and Methods of Electrical Measurement, Izdat. Sibirskoe Otdelenie A. N. SSSR, Novosibirsk, 1966.

1094 G. F. Verzakov, et al., "Introduction to Engineering Diagnostics", Izdat. Energiya, Moscow 1968.

1095 J. Vignolle, "Decoupage Logique D'un Systeme en Vue de Faciliter la Simulation et ses Tests", Contrat Cri 70025, April 1971.

1096 J. Vincent-Carrefour, "Controle et Depannage D'Un Ensemble Logique", Convention DGRST 67-00966.

1097 J. Vincent-Carrefour, "Etude de L'Influence des Aleas de Fonctionnement D'Une

Carte Logique sur une Sequence de Test", Rapport Final, Convention CRI 70023, 1972.

1098 S. R. Vishnubhotla and Y. H. Chuang, "An Analysis Approach to the Diagnosis of Combinational Circuits", Computer System Lab., Washington University, Technical Memo No. 119, April 1971.

1099 S. R. Vishnubhotla and Y. H. Chuang, "A Path Analysis Approach to the Diagnosis of Combinational Circuits", 8th Design Automation Workshop, 1971.

1100 R. E. Vogelsberg, "Identifying Sources of Unknown Levels Generated During Three-Value Fault Simulation", *IBM Technical Disclosure Bulletin*, Vol. 14, No. 8, p. 2510-12, January 1972.

1101 R. E. Vogelsberg, "Optimization of Fault Simulation by Pre-Analysis", *IBM Technical Disclosure Bulletin*, Vol. 14, No. 2, p. 2508-9, January 1972.

1102 A. F. Volkov, V. A. Vedeshenkov, and V. D. Zenking, "Automatic Failure Diagnosis in Control Computers", Third Congress of the International Federation of Automatic Control, June 20-25, 1966.

1103 R. B. Walford, "The Lamp System," Workshop on Fault Detection and Diagnosis in Digital Circuits and Systems, Lehigh University, 7-9 December 1970, p. 10.

1104 R. M. Walker and W. C. Jensen, "The Mythology of LSI Functional Test Generation", *Nerem 1970*, p. 94-95.

1105 R. L. Walquist, "The Isolation and Correction of Computer Malfunctions", AD-609 276, September 1965.

1106 K. C. Wang, "Design of Asynchronous Sequential Machines with Fault-Detection", Proceedings of the Allerton Conference 1969, p. 237-42.

1107 G. C. Warburton, "Automatic Dynamic Response System for Testing Semiconductors", Joint Conference on Automatic Systems, IEEE Conference Proceedings No. 17, p. 467-84, April 1970.

1108 R. L. Ward and T. O. Holtey, "The Maintainability Factor in the Design of Digital Systems Using Microelectronics", *Wescon 1966*, 25-5, p. 1-7.

1109 E. N. Warshawsky, "An Introduction to Computer Diagnostics", *Computer Design*, July 1966, p. 58-61.

1110 G. B. Way and M. Rubin, "Automatic System Testing", *Automation*, Vol. 6, p. 72-78, September 1959.

1111 C. D. Weiss, "Cell Input-Set Systems and Their Use in Fault Detection for Combinational Networks, Pt. 1", Fault Detection Tests for Cell Input-Set Systems, January 1971.

1112 C. D. Weiss, "Bounds on the Length of Terminal Stuck-Fault Tests", *IEEE Transactions on Computers*, Vol. C-21, No. 3, p. 305-09, March 1972.

1113 A. S. Weitzenfeld and W. W. Happ, "Combinatorial Algorithms for Computer-Oriented Fault Identification in Multi-Terminal Devices", Proceedings of the IEEE-ESD Symposium, April 1967.

1114 G. Wend, R. A. Lloyd, and T. A. Keller, "Large Array Test Program Generator", *Automatic Support System for Advanced Maintainability*, 1970, p. 148-152.

1115 G. E. Whitney, "Algebraic Fault Analysis for Constrained Combinational Networks", *IEEE Transactions on Computers*, Vol. C-20, No. 2, p. 141-48, February 1971.

1116 G. E. Whitney, "Test Generation for Structural Two-State Machines by a Method of Sequential Constraints", 1971 IEEE International Computer Society Conference, p. 73-74.

1117 W. R. Wilcox and S. P. Lapage, "Automatic Construction of Diagnostic Tables", *The Computer Journal*, Vol. 15, No. 3, p. 263-67, October 1972.

1118 J. C. Wilcox, "Competitive Evaluation of Failure Detection Algorithms for Strapdown Redundant Inertial Instruments", N73-22608, April 1973.

1119 G. H. Williams and W. G. Zavisca, "Comments on Sequential Machine Identification", *IEEE Transactions on Computers*, Vol. C-21, No. 6, p. 616, June 1972, Voir J. Kella, IEEE Trans. on Computers, Vol. C-20, No. 3, p. 323-328, March 1971.

1120 L. R. Williams, "Automating the Gap Between Logic Simulation and Testing", *Automatic Support System for Advanced Maintainability*, 1970, pp. 142-147.

1121 L. R. Williams, "Automated Digital Test Pattern Generation", Semiconductor Integrated Circuit Processing and Production Conference, Anaheim, 9-11 February 1971, p. 322-28.

1122 M. J. Y. Williams, "Test Generation for LSI Sequential Circuits, The Use of Test Points in Conjunction with Additional Logic", Ph.D. Dissertation, Department of Electrical Engineering, Stanford University, August 1970.

1123 M. J. Y. Williams, "Enhancing Testability of Large-Scale Integrated Circuits Via Test Points and Additional Logic", AD-714-511, September 1971.

1124 M. J. Y. Williams and J. B. Angell, "Enhancing Testability of Large-Scale Integrated Circuits Via Test Points and Additional Logic", *IEEE Transactions on Computers*, Vol. C-22, No. 1, pp. 46-60, January 1973.

1125 B. B. Winter, "Optimal Diagnostic Proceedings", *IRE Transactions on Reliability and Quality Control*, Vol. RQC-9, p. 13-19, December 1960.

1126 R. D. Wollmer, "Selecting and Sequencing Tests in an Adaptive Countdown," 30th National Meeting of Operations Research, 1966.

1127 C. P. Womack, "Schmoo Plot Analysis of Coincident-Current Memory Systems", *IEEE Transactions on Electronic Computers*, Vol. EC-14, p. 36-44, February 1965.

1128 W. H. Woodruff, "A Computer Controlled Test Facility for Film Memory Arrays", *IEEE Transactions on Magnetics*, Vol. MAG-6, No. 3, p. 579-80, September 1970.

1129 R. D. Wooster, "Techniques of Single and Multiple Fault Analysis of MSI Digital Logic Arrays", *Intercon 1972*, p. 542-3.

1130 A. S. Wyllie, "Diagnostic Techniques Applied to Fault Isolation Within a Programmable Communications Controller", International Conference on Communications, June 11-13, 1973.

1131 S. V. Yablonskii and I. A. Chegis, "Tests for Electrical Networks", *Uspekhi Matematicheskikh Nauk*, Vol. 10, No. 4, p. 182-4, 1955.

1132 E. A. Yacubaytis, "Checking Tests of Combinational Finite Automata", *Automation and Computer Engineering*, 1968, No. 3, p. 13-20.

1133 V. F. Yadiha, "An Algorithm for Determining Diagnostic Functions for Checking Tests for a Combinated Automaton", in *Problems of Synthesis of Finite Automata*, V. I. Levin Ed., p. 161-70, Riga, USSR, Izdatel'Stvo 'Zinatne', 1972.

1134 H. Yamamoto, T. Watanabe, and Y. Urano, "Alternating Logic and Its Application to Fault Detection".

1135 M. Yamamoto and K. Kinoshita, "Modular Structured Circuits and Their Fault Diagnoses", Technol. Rep. Osaka Univ., Vol. 21, No. 995-1026, p. 599-612, October 1971.

1136 S. S. Yau and C. K. Tang, "Universal Logic Circuits and Their Modular Realizations", SJCC 1968, p. 297-305.

1137 S. S. Yau and C. K. Tang, "Universal Logic Modules and Their Applications", IEEE Transactions on Computers, Vol. C-19, No. 2, p. 141-49, February 1970.

1138 S. S. Yau and M. Orsic, "Fault Diagnosis and Repair of Cutpoint Cellular Arrays", IEEE Transactions on Computers, Vol. C-19, No. 3, p. 259-62, March 1970.

1139 S. S. Yau and Y. H. Tang, "An Efficient Algorithm for Generating Complete Test Sets for Combinational Logic Circuits", IEEE Transactions on Computers, Vol. C-20, No. 11, p. 1245-51, November 1971.

1140 T. Yeager, "Automation of Test Specification for N/C Printed Circuits Boards", Seventh Annual Design Automation Workshop, 1970.

1141 Y. T. Yen, "Intermittent Failure Problems of Four-Phase MOS Circuits", IEEE Journal of Solid State Circuits, Vol. SC-4, No. 3, p. 107-110, June 1969.

1142 Y. T. Yen, "A Method of Automatic Fault Detection Test Generation for Four-Phase MOS LSI Circuits", SJCC 1969, p. 215-20.

1143 Y. T. Yen, "Computer-Aided Test Generation for Four-Phase MOS LSI Circuits", IEEE Transactions on Computers, Vol. C-18, No. 10, p. 890-4, October 1969.

1144 V. A. Yermilov and N. I. Kudyukov, "Fault Diagnosis in Digital Systems", Cybernetics Engineering, Vol. 8, No. 3, p. 553-57, May-June 1970.

1145 I. H. Yetter, "High Speed Fault Simulation for Univac 1107 Computer System", Proceedings of the 23rd National Conference of the ACM, Publication p. 68, 265-77, 1968.

1146 I. H. Yetter, "Fault Isolation Dictionary for the Univac 9200/9300", Computers Designer's Conference, Anaheim, California, 19-21 January 1971, pp. 433-443.

1147 M. Yoeli and S. Rinon, "Application of Ternary Algebra to the Study of Static Hazards", Journal of the ACM, Vol. 11, pp. 84-97, January 1964.

1148 K. K. Yoshikazu and Y. Kasahara, "Diagnosis of Multiple Faults in Combinational Circuits", Electronics and Communications in Japan, Vol. 52-C, No. 4, p. 123-31, 1969.

1149 F. M. Young, "An Innovation Approach to the Testing of MOS on the Production Line", Wescon Technical Papers, 1971.

1150 H. W. Young, "Specifying the Interface Between Design and Test Organizations", Joint Conference on Automatic Test Systems, IEEE Conference Proceedings No. 17, April 1970, p. 91-110.

1151 E. G. Zavisca and G. H. Williams, "An Approach to Sequential Machine Identification", Proc. IEEE Region 3, 1972.

1152 D. B. Zeh, "A MOS LSI Computer ANS System, Logic and Circuit Design Verification and Testing", Proceedings of the IEEE Naecon 1969, p. 447-52.

1153 Y. A. Zholkov, "A Method for Testing Combinational Logic Circuits", Trudy Inst. Nii Teplopribor, No. 1, 1964.

1154 Y. A. Zholkov, "On the Simplification of an Algorithm for Minimizing the Numbers of Tested States in One Method for Testing the Elements of

Combinational Circuits", Trudy Inst. Nii Teplopribor, No. 3, 1964.

1155 Periodicals
Vsesoyuznoe Soveshchanie po Tekhnicheskoi Diagnostike
Trud.IEEE Transactions on Computers.

1156 Anonyme, "Put Your Memory to a Test", *The Electronic Engineer*, August 1970,
p. 49.

1157 Battelle Memorial Institute, "The Formulation of Automatic Check-Out Techniques", ASD Technical Documentation Report No. 33-616-7761, March 1962.

1158 "Computer Testing Simplified," *Computer Design*, February 1971, p. 24, 26.

1159 Aerospace Research Applications Center, "The Evaluation of Failure and Failure
Related Data", ARAC 1972.

1160 Aries Corp., "Computer Diagnosis Study", NASA Goddard Space Flight Center,
Greenbeld, Maryland, Report on Contract No. NASS-3147, January-July 1963.

1161 Automation Analysis Inc., "DFA System (Digital Fault Analysis)."

1162 Computer Automation Inc., "Software Packages Optimize Use of Automatic Card
Testers", *Computer Design*, November 1971, p. 30.

1163 NASA, "DALG-A Program for Test Pattern Generation in Combinational Logical
Circuits", N72-12117.

1164 Pacific Applied Systems, Inc.—TASC System, Voir *Computer Design*, November
1971, p. 26.

1165 Proceedings of the 1971 IEEE International Symposium on Fault-Tolerant
Computing, IEEE Publication No. 71C6-C; proceedings of the 1972 IEEE
International Symposium on Fault-Tolerant Computing, IEEE Publication No.
72CH0623-9C; proceedings of the 1973 IEEE International Symposium on
Fault-Tolerant Computing, IEEE Publication No. 73CH0772-4C; First Workshop
on Fault Detection and Diagnosis in Digital Circuits and Systems, Lehigh
University, December 7-9, 1970.

1166 Second Workshop on Fault Detection and Diagnosis in Digital Circuits and
Systems, Lehigh University, December 6-8, 1971.

1167 Sesa, "Localisation des Defauts dans une Logique Combinatoire", Synchrone ou
Asynchrone Synchronisee, Contrat DRME No. 70491, September 14, 1971.

1168 Siemens A-G., "Fault Diagnosis in Stored-Program Modular Systems", *Elektron-
ische Rechenanlagen*, Vol. 12, pp. 150-154, June 1970.

1169 "Techniques of Digital Troubleshooting", Hewlett Packard Application Note
163-1.

1170 "Technitrol Delivers. . . ," *Computer Design*, February 1971, p. 24.

1171 Telpar Inc., "Fault Isolation and Diagnostic Test Techniques for Large Arrays in
Sequential Digital Networks", Program Plan N.7116432.

1172 Vought Aeronautics and Co., "Lasa (Logic Automated Stimulus and Response,"
Voir Electronics.

1173 Westinghouse Electric Corp., "Comtest System."

1174 "Hardware Testing of Digital Process Computers, Recommended Practice",
Instrument Society of America, 1972.

1175 IEE Conference on the Automation of Testing, The University of Keele,

Stoke-on-Trent, Angleterre, 20-22 September 1972.

1176 IERE Conference on Automatic Test Systems, University of Birmingham,England,
      April 14-17, 1970.

1177 *IEEE Transactions on Computers,* Vol. C-20, No. 11,   Special Issue on Fault-
      Tolerant Computing.

1178 *IEEE Transactions on Computers,* Vol. C-22, No. 3, Special Issue on Fault-
      tolerant Computing.

1179 MOS Array Testing, pp. 365-373 in MOS Integrated Circuits by the Engineering
      Staff of American Microsystems Inc., Van Nostrand, New York, 1972.

1180 MOS LSI Test Considerations, pp. 321-330 In MOS Instegrated Circuits by the
      Engineering Staff of American Microsystems Inc., Van Nostrand, New York, 1972.

INDEX